System-Level Design Methodologies for Telecommunication

Nicolas Sklavos • Michael Hübner
Diana Goehringer • Paris Kitsos
Editors

System-Level Design Methodologies for Telecommunication

Springer

Editors
Nicolas Sklavos
Technological Ed Institute of
Patras
Hellas
Greece

Diana Goehringer
Ruhr-Universität Bochum
Bochum
Germany

Michael Hübner
Ruhr-Universität Bochum
Bochum
Germany

Paris Kitsos
Technological Educational Inst of Patras,
Informatics & MM
KNOSSOSnet Research Group
Pyrgos
Greece

ISBN 978-3-319-37593-9 ISBN 978-3-319-00663-5 (eBook)
DOI 10.1007/978-3-319-00663-5
Springer Dordrecht Heidelberg London NewYork

Softcover reprint of the hardcover 1st edition 2014

Printed on acid-free paper

Springer is part of Springer Science+Business Media (www.springer.com)

Contents

Contributors

D. Amaxilatis Computer Technology Institute and Press “Diofantus”, Patras, Greece

George Asimakopoulos Department of Telecommunication Systems and Networks (TESYD), Technological Educational Institute of Messolonghi, Hellas, Greece
e-mail: asim@teimes.gr

Labros Bisdounis Electrical Engineering Department, Technological Educational Institute of Patras, 1, M. Alexandrou Street, 263 34 Patras, Greece
e-mail: bisdounis@teipat.gr

David Fuschelberger Department of Telecommunication Systems and Networks (TESYD), Technological Educational Institute of Messolonghi, Messolonghi, Greece
e-mail: d.fuschelberger@gmail.com

Apostolos P. Fournaris Electrical and Computer Engineering Department, University of Patras, Patras, Greece
e-mail: apofour@ieee.org

Angelina Gkioni Department of Telecommunication Systems and Networks (TESYD), Technological Educational Institute of Messolonghi, Hellas, Greece
e-mail: angegkio@gmail.com

L. Gortzis University of Patras, 26504 Rio, Greece

V. Georgitzikis Computer Technology Institute and Press “Diofantus”, Patras, Greece
e-mail: georgitzik@ceid.upatras.gr

Georgios Keramidas Electrical and Computer Engineering Department, University of Patras, Patras, Greece
e-mail: keramidas@ece.upatras.gr

Spiros Louvros Department of Telecommunication Systems and Networks (TESYD), Technological Educational Institute of Messolonghi, Hellas, Greece
e-mail: splouvros@gmail.com

Fotis Plessas Department of Telecommunication Systems and Networks, Technological Education Institute of Messolonghi, Messolonghi, Greece
e-mail: fotis.plessas@gmail.com

Aggeliki Pragiati Hellenic Telecommunications & Post Commission (EETT), Athens, Greece
e-mail: apragiati@eett.gr

Ch. Pylarinou University of Patras, 26504 Rio, Greece

Vassilios Triantafyllou Department of Telecommunication Systems and Networks (TESYD), Technological Educational Institute of Messolonghi, Hellas, Greece
e-mail: triantaf@teimes.gr

Nikolaos Terzopoulos Department of Computing & Communication Technologies, Faculty of Technology, Design & Environment, Oxford Brookes University, Wheatley, Oxford, UK

S. Zimeras University of the Aegean, Karlovassi, 83200 Samos, Greece
e-mail: zimste@aegean.gr

Chapter 1
Indoor Radio Design: LTE Perspective

Spiros Louvros, Vassilios Triantafyllou and George Asimakopoulos

Abstract Long-term evolution (LTE) indoor coverage is becoming important day by day due to multilayer design and high traffic-building premises. Nowadays, it is true that user expectations from operator's indoor high-quality services and capacity availability provides a well-promised opportunity to offer improved LTE services with appropriate traffic revenues. Customers expect indoor faster Internet connections than ever and they would not tolerate slow connections. Wireless network indoor capacity for data services will become more important in the near future. As a result, indoor LTE radio design is one of the key elements to provide a high-quality service to meet the user demand for a high-capacity mobile network.

1.1 Introduction

This chapter will provide a complete radio design perspective for indoor long-term evolution (LTE) services supporting all available mobile network standards. In order to have a complete and interference-free design, indoor antenna is important. This antenna could be used for indoor services over all mobile standards (LTE, High Speed × Packet Access (HSxPA+), wideband code division multiple access (WCDMA), and global system for mobile communications (GSM)) together with nonmobile standards (wireless local area network (WLAN)). Radio sharing could also be the case where same infrastructure could be used by several operators. In the same way, a total new radio deployment idea could also be used in order to have a completely interference-free radio environment; this is the use of visible light communication (VLC) also known as WiLi solution. In such a solution, all antenna infrastructure is replaced by power light-emitting diodes (LED) lights and indoor LTE coverage

S. Louvros (✉) · V. Triantafyllou · G. Asimakopoulos
Department of Telecommunication Systems and Networks (TESYD),
Technological Educational Institute of Messolonghi, Hellas, Greece
e-mail: splouvros@gmail.com

V. Triantafyllou
e-mail: triantaf@teimes.gr

G. Asimakopoulos
e-mail: asim@teimes.gr

N. Sklavos et al. (eds.), *System-Level Design Methodologies for Telecommunication*,
DOI 10.1007/978-3-319-00663-5_1, © Springer International Publishing Switzerland 2014

is provided solely by light sources. Generally speaking, indoor radio design might be possible for new indoor projects, design of single-operator indoor antenna infrastructure, multioperator indoor design, or multinetwork design including LTE, WCDMA, GSM, and/or WLAN. From design process point of view, indoor radio design might be similar to the macro-cell outdoor design; however, some specific aspects must be emphasized. Of course, site acquisition is not needed where indoor coverage objective is clearly identified.

Proposed indoor process or technique selection depends on coverage and capacity requirements definitions. Proposed solution thus is selected based on the signal strength and signal quality targets, number of indoor walls and construction material, number of floors, floor layout, and dimensions. In such design process, the decisions concerning the type of the distribution network and the number of antenna heads are also important. The positions of the antennas are planned considering practicable solutions and the results of prediction calculations. When preliminary indoor design is completed, building details should be verified and all installation concerns and facility availabilities should be considered carefully. Practical limitations should also be considered into proposed planned configuration and the use of measurement tools should be proposed, evaluating and approving coverage and interaction (interferences) between planned system and existing macro-cell outdoor coverage. After appropriate installations, planning issues and antennas' location selection, signal strength loss distribution should be measured separately for each antenna resulting into required indoor transmission output power. Last concern should be the indoor network capacity, showing explicitly the needs on number of cells, antenna tilts, cell directivity, and sectorization.

1.2 Preliminary LTE Indoor Design Requirements

When planning for indoor coverage and services, it is important to realize that a distributed antenna [1] system will have to be used for micro, pico, or femto cells, as presented in Fig. 1.1. European-Union (EU) and 3G Patent Platform (3GPP) standards have to be followed and all spectrum requirements should be designed fulfilling and respecting these standards. Antennas and radio design proposals should be compliant with current baseline and future technologies plus the operating band requirements. Moreover, all passive components (feeder systems, combiners, splitters, and antennas' equipments) should be compliant with future technologies such as visible light communications (VLC). Coverage expectations provide all necessary restrictions for primary radio-frequency (RF) design criteria and are defined by the respective reference signal strength provided most often by the operator. Specifically from indoor perspective, indoor coverage is typically defined by both downlink and uplink signal strength and vary by access technology layer for corporate and public sites.

To overcome any future expected problems on accessibility or signal strength coverage, it is recommended that all indoor antenna-distributed networks should be planned using multiple sectors. For each sector separate planning should follow

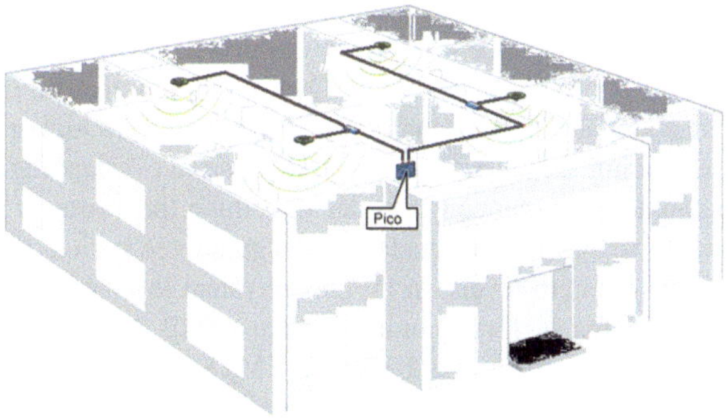

Fig. 1.1 Indoor coverage example, using pico-cell antennas

GSM 900, Digital Cellular Service (DCS) 1800, Universal Mobile Telecommunication System (UMTS) 2100, LTE 2600 MHz, and WiFi frequency bands and number of sectors and antenna branches will be dependent on both the architectural constraints as well as the Single Input Single Output/Multiple Input Multiple Output (SISO/MIMO) capacity requirements. During planning process, data services should be guaranteed [2]. Consequently, a general rule of thumb might be that signal strength level of − 90 dBm in 90–95 % of coverage area should be provided with adequate carrier-to-interference (C/I) level of minimum 9 dB. Worst radio conditions will result into very low throughput and low-service integrity. As an example from real drive tests, orthogonal frequency-division multiplexing (OFDM) resource block (RB) throughput versus signal-to-noise and interference ratio (SINR) curve is provided in Fig. 1.2. It is obvious that as long as SINR is kept low because of bad indoor planning, expected throughput per RB falls below 10 kbps [3].

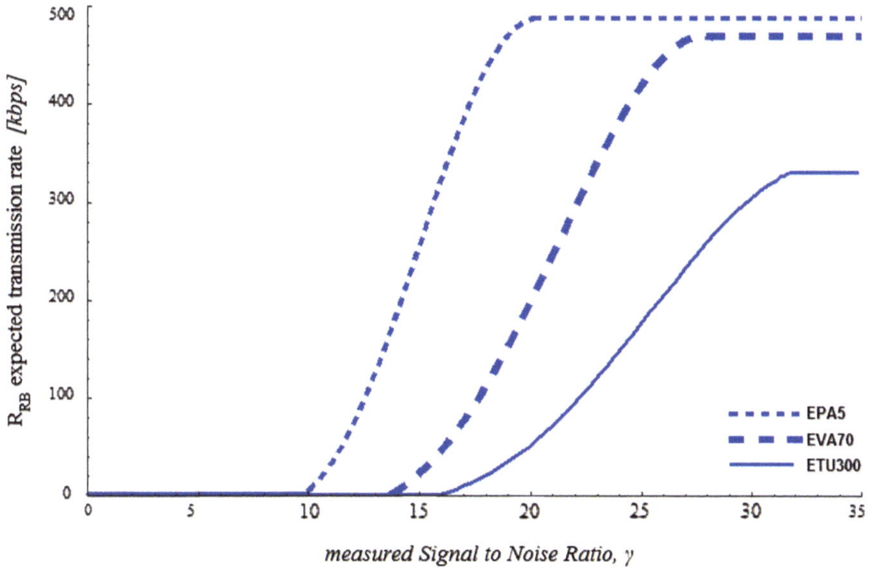

Fig. 1.2 Expected transmission rate per RB versus signal-to-noise ratio γ

Specifically for LTE planning, certain requirements should be fulfilled [3]. First of all, the reference signal received power (RSRP) should be better than 85 dBm in 90–95 % of indoor coverage. Quality should also be guaranteed, so that the reference signal received quality (RSRQ) should be higher than – 10 dB. Specifically, when multisector design is preferred and in overlapping locations, server dominance should be guaranteed in order to reduce the waste of network resources and network performance degradation. To guarantee such a criterion reference signal (RS), signal strength levels on OFDM physical layer of dominant server, among several overlapping sectors, should be at least 6 dB stronger.

When coexistence of LTE with UMTS/HSxPA is a case, then the following criteria should be fulfilled. First of all downlink CPICH_RSCP power should be better than – 80 dBm in 90–95 % of indoor environment. Moreover, quality on CPICH_Ec/No should be higher than – 10 dB. Dominance with one strong server in overlapping locations should also be guaranteed. In 3G systems, dominance is a function of pilot pollution, a reference measurement on degradation in CPICH_Ec/No of best pilot server in the presence of other pilot signals. As a general case, in locations where there are more than three strong servers, pilot pollution is considered to be bad. Whenever only two dominant servers exist in the desired planned location, CPICH_Ec/No of desired dominant server should be 5 dB stronger than CPICH_Ec/No of neighbor server.

1.3 Planning Area Surveys

Planning prerequisites [2] have to be followed by indoor surveys. Preliminary design will have expected antenna locations, cabling, and assumed equipment locations. Sometimes repeaters might be needed, consequently extra planning requirements and equipment locations should be guaranteed. Signal strength confirmation is evaluated or estimated either by system simulations or by real drive tests using a test transmitter and receiver per expected antenna locations. Whenever repeaters are used, RSSI for GSM/LTE or Pilot Power for WCDMA at the donor antenna should be measured on the roof. Specifically for GSM solutions, broadcast control channel (BCCH) frequency along with the base station identity code (BSIC) should also be identified. Moreover, outdoor-to-indoor coverage in the indoor area from surrounding sites should be measured and reported.

To avoid coexistence interradio technology interference, a detailed outdoor-to-indoor planning report should be prepared regarding all available BCCH with BSIC as presented in Fig. 1.3. This will help to determine the design thresholds required as well as identify the handover locations.

Sometimes, the expected antenna system and site/sector locations might not be adequate. In such scenario, planners should have an agreement with the operator regarding possible alternative locations for cabling, mounted distributed antenna system, or communication E-UTRAN Node B (eNodeB) equipment. After final location surveys, required locations of all the distributed antennas on each individual

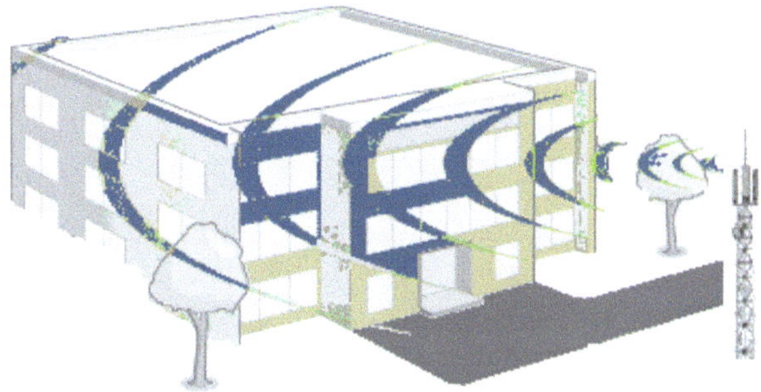

Fig. 1.3 Expected transmission rate per RB versus signal-to-noise ratio γ

floor should be decided. Information regarding building characteristics should be acquired for future installations as shown in Fig. 1.4.

Ceiling type is important indicating possible areas requiring special attention due to decorations, sensitive areas, etc. Impenetrable obstacles should be explicitly indicated, specifically for locations with supporting building walls, steel concrete partitions, etc.

1.4 LTE Indoor Coverage Design

The most common proposed and preferred solution for indoor coverage design is the use of a number of dedicated base stations as presented in Fig. 1.5 [4]. It is in general a solution with good performance results for indoor coverage on large buildings,

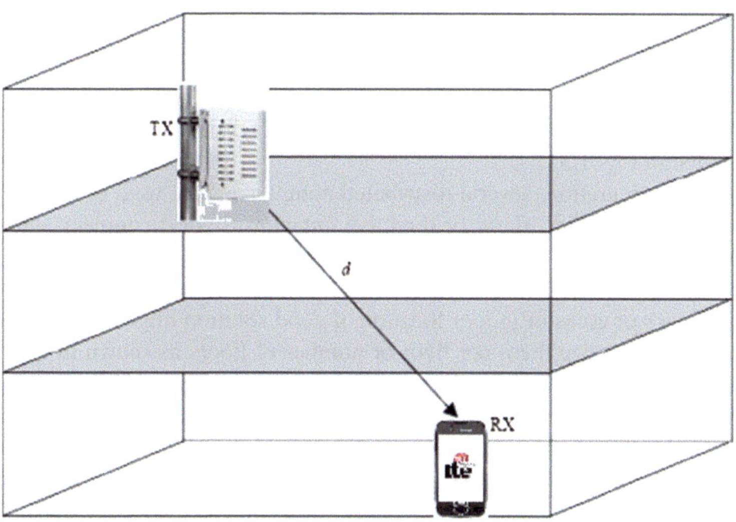

Fig. 1.4 Buildings characteristics for indoor planning

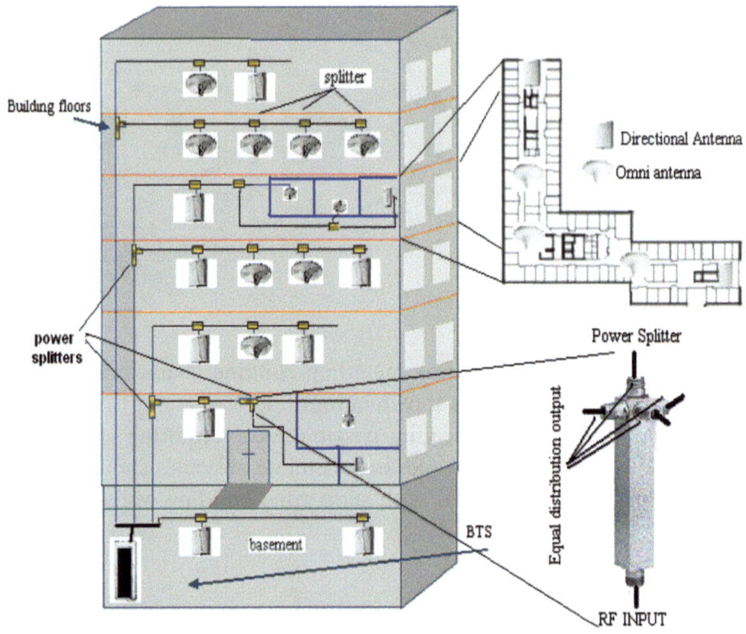

Fig. 1.5 Indoor coverage with number of distributed antennas

malls, airports, university campuses, and metro business district areas. It is supposed to be a good solution since nominal planning is performed, something that most cell planners already know very well, where both coverage and capacity are fulfilled. Moreover, interference analysis is successfully defined and signal strength received levels are confirmed. This is a solution quite often radio designers prefer in order to optimize multilayer network coverage design by optimizing indoor coverage and offloading capacity of existing indoor-to-outdoor macro cells. General rule of thumb [1] is to connect one eNodeB sector output power to a distributed system of antennas as presented in Fig. 1.5.

In such configuration, several distributed components are used extensively such as coaxial feeder cables, omni or directive antennas, hybrid couplers, and power splitters.

However, in order to increase capacity, if required because of extensive traffic load on some floors or areas of indoor location, a good solution might be to use several eNodeBs and sectorize them per floor or number of floors as shown in Fig. 1.6. If this is the case, the implementation cost is increased; however, the capacity profit overcomes any other disadvantage.

Specifically for tunnels or underground metro tubes [1], another solution might be the leaky cables (a specific design of radiating feeders). The main disadvantages of such a solution are high losses and short range. Indeed, losses are dependent on carrier frequency and coverage range is limited longitudinal to few hundred meters

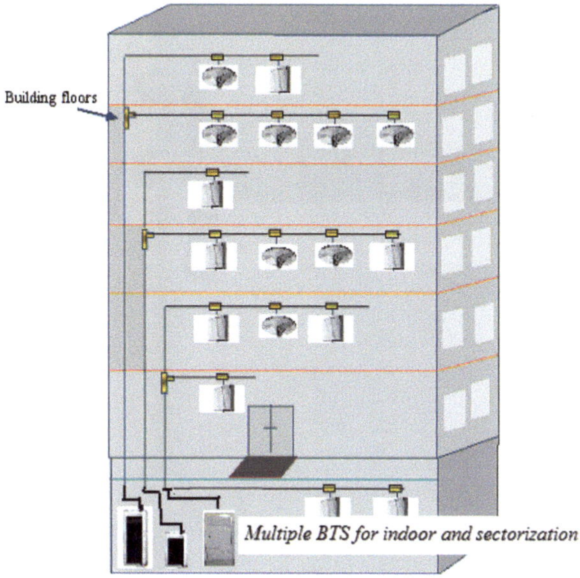

Fig. 1.6 Indoor coverage with use of multiple eNodeBs

and transversal to few ten meters. This is however adequate for most of the tunnel or underground applications. In special cases of longer distance coverage, leaky cables are not recommended and active antenna-distributed systems are rather preferred and standardized, see Fig. 1.7. eNodeB is split into two subsystems, the base band unit (BB, responsible for all base band functions such as modulation, coding, and signal processing) and the radio remote unit (RR, responsible for the radio functionality as power amplification and filtering). Splitting is important in order to place BB and RR unit in different locations. BB unit could be placed close to the main base station equipment and the RR unit side by side to the antenna system inside the tunnel. In the past, most of the BB units were connected to RR units through feeders; however, losses were high and either repeaters or high initial power was needed. Nowadays,

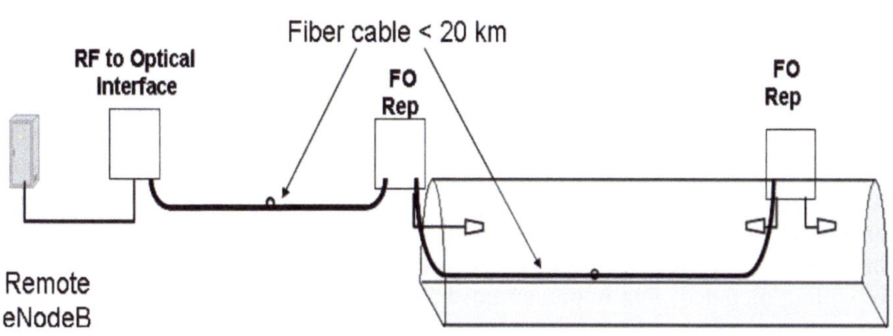

Fig. 1.7 Indoor coverage using fiber optics and repeaters

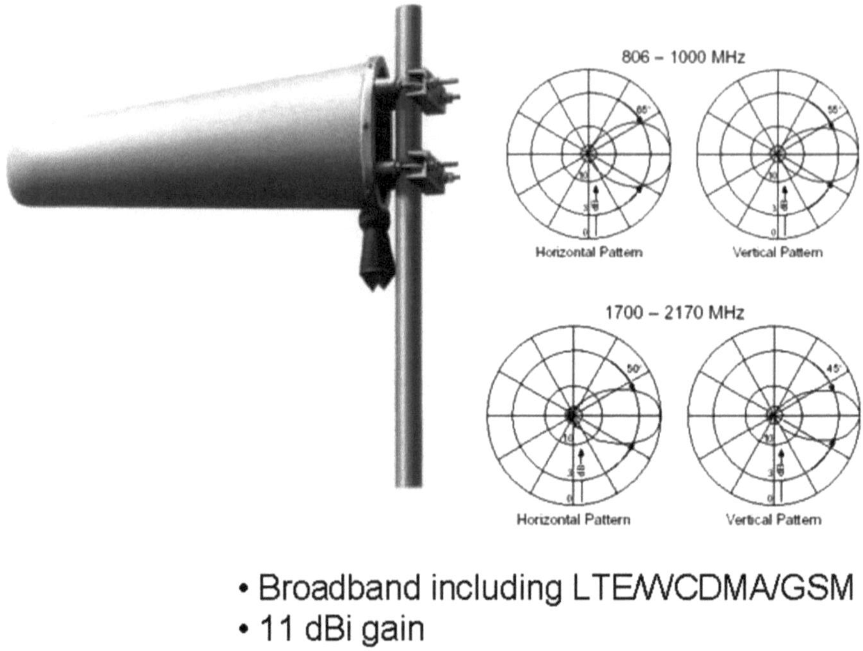

Fig. 1.8 Indoor in-tunnel coverage antenna type with its radiation characteristics

however, the preferred solution is the RF to optical transmission with the use of optical fibers in order to eliminate passive losses.

Types of antennas [5] are extremely important in such scenario since directivity and gain are very crucial. In Fig. 1.8, a typical antenna with its radiation characteristics is presented.

Fiber optics [1] could also replace main transport domains for metro and indoor planning when large buildings are involved or when multiple buildings are to be covered, mainly in metropolitan areas, as presented in Fig. 1.9. This is extremely recommended since optical fiber infrastructure is already installed and available.

According to Fig. 1.10, cell planners should always remember that outdoor-to-indoor interference should be kept at low levels.

Another important parameter is the antenna directivity [5] and radiation diagram in indoor planning cases. Indeed, for indoor coverage directional antennas should be used depending on the indoor topology. However, due to radiation patterns there are areas of low coverage even though user is close to the antenna. For directional antennas with twin lobes or omni antennas, low-gain area is right below the antenna; hence planners should avoid placing the antenna in such a location (wall mounted or ceiling mounted) that important coverage indoor spaces suffer from low gain. According to Figs. 1.11 and 1.12, on the other hand when directional antenna with on lobe is used [4], the low-gain area is on the diagonal of the main lobe.

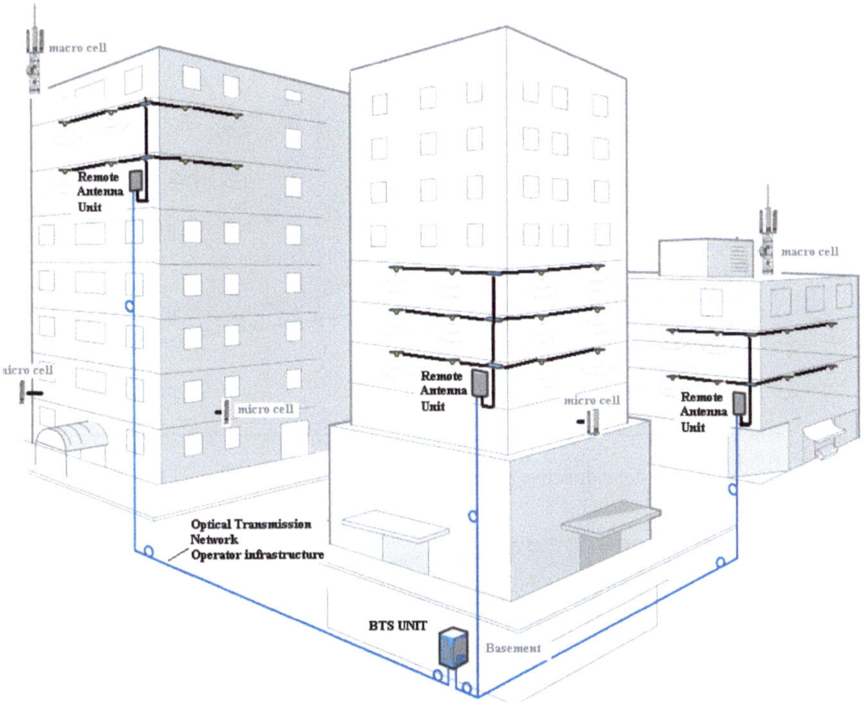

Fig. 1.9 Indoor coverage with fiber optics metropolitan transport network

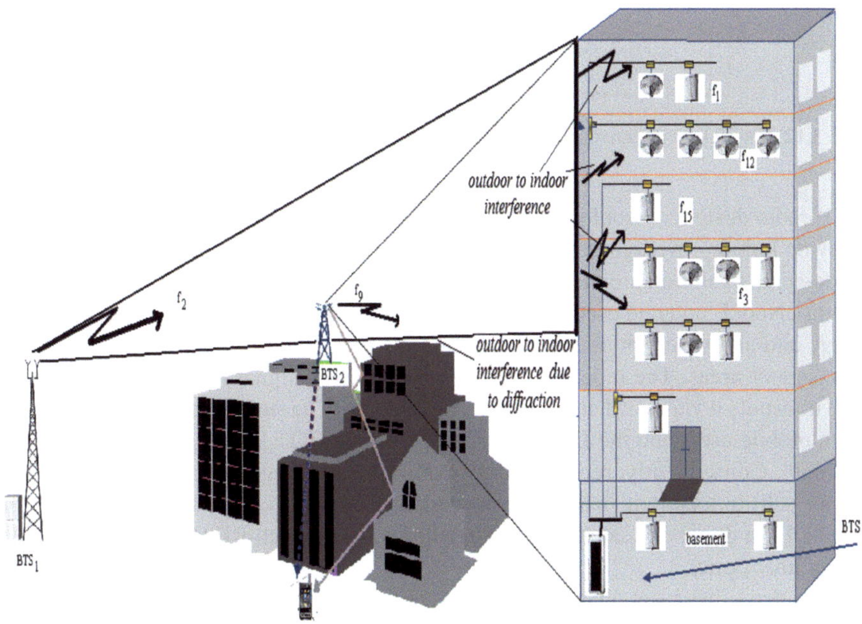

Fig. 1.10 Outdoor-to-indoor potential interference

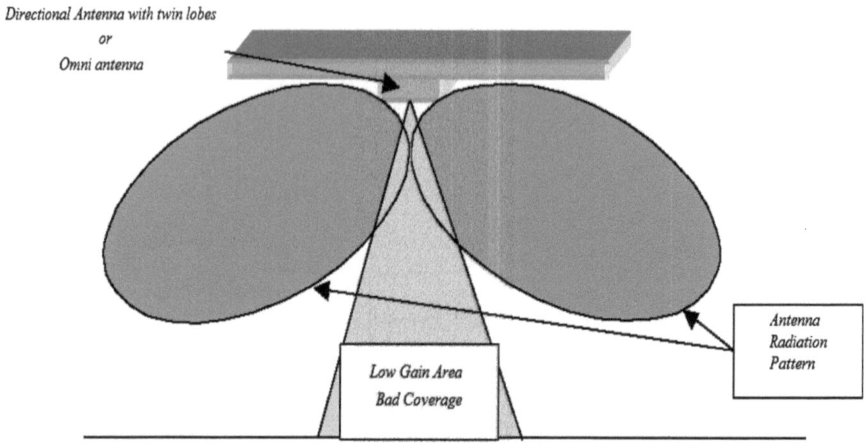

Fig. 1.11 Omni- or twin-lobe directive antenna low-gain coverage area

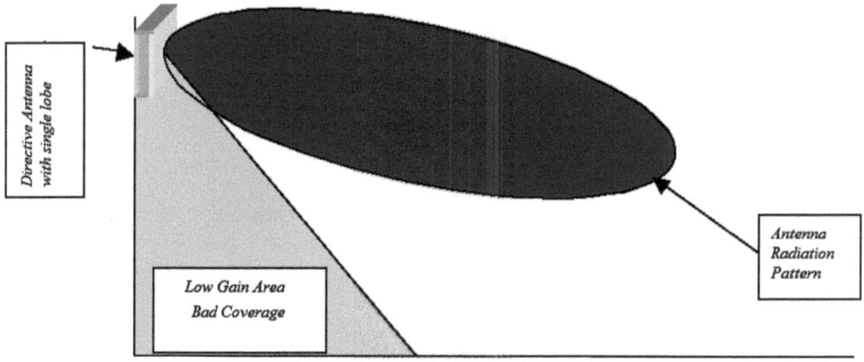

Fig. 1.12 Single-lobe directive antenna low-gain coverage area

If radio designers would not like to use optical fibers or leaky cables in order to provide indoor coverage in tunnels or underground areas, another well-known and wide-spread solution are the radio repeaters [1]. Radio repeaters rely on an outdoor donor eNodeB antenna used for outdoor communication with existing outdoor coverage and an indoor service antenna that aims to extend original outdoor coverage into indoor areas. The repeater amplifies the received signal from outdoor antenna and transmits it via the other indoor service antenna. The signal amplification enables both mobile users and the Base Transceiver Station (BTS) to receive a better signal strength. Consequently, repeaters only extend the coverage of a particular cell (the donor cell) to locations that were not originally part of the coverage of that cell. In Fig. 1.13, a donor–repeater pair is presented together with RF to optical interfaces using fiber optics.

The main disadvantage of using repeaters is the capacity overloading since they do not provide any additional capacity and on the same time they do not allow offloading the macro cell from which the donor antenna picks up the signals. As

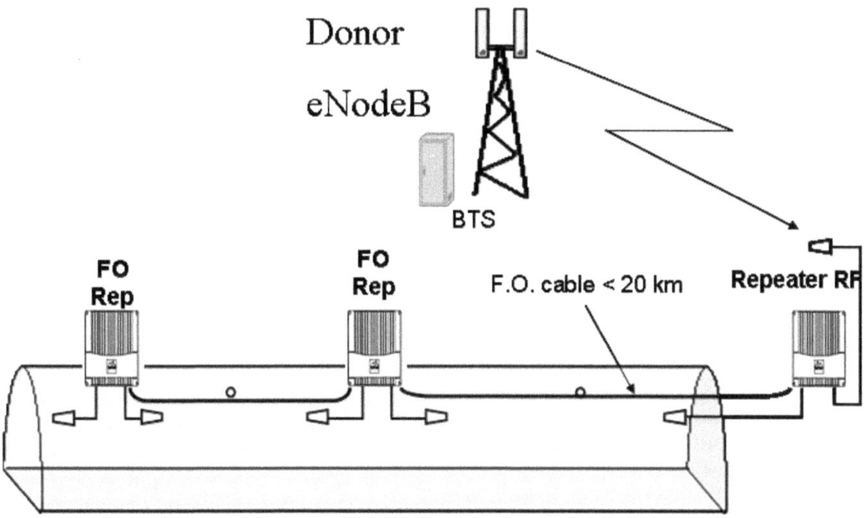

Fig. 1.13 Donor–repeater indoor—in-tunnel coverage scenario

a rule of thumb: whenever RF repeaters are used, capacity planning of outdoor donor cells should be reconsidered and increased, otherwise traffic congestion and performance degradation might be expected. Another disadvantage is interference. Indeed, when extending the outdoor coverage into indoor areas, it might disturb the frequency planning by extending specific carriers into unwanted areas outside in-building coverage; this is something planner should always keep in mind. The main advantage of using repeater is the low implementation cost, the use of regular transmission equipment, and the enhancement of indoor coverage with fairly easy deployment. From coverage perspective, coverage improvement may be assured in a short time with low cost.

According to Fig. 1.14, another possible solution [1] might be the use of leaky cabling as the serving antenna.

Generally speaking, radio repeaters are widely used for indoor coverage expansion and optimization. Indoor expansions using repeaters are widely used when external outdoor network is not able to provide a satisfactory service in certain conditions. Specifically for data services on WCDMA or LTE coverage scenarios, where low SINR is the dominant case, repeaters are recommended. Moreover, in dense urban areas with high building factor, lower floors usually suffer from low signal strength due to penetration pat losses. Repeaters are quite often proposed and used to expand coverage. Also, fast data delivery service depends mostly on signal levels and on medium access control (MAC) scheduler decisions. Expanding indoor coverage improves capacity and throughput performance, especially for high-speed packet access (HSPA), evolved high-speed packet access (HSPA$^+$), and LTE. Nowadays, they are used for in-building coverage expansions, underground expansions, in-tunnel coverage, in-train coverage expansions, in-airplane expansions, and of course for near-cost ferries.

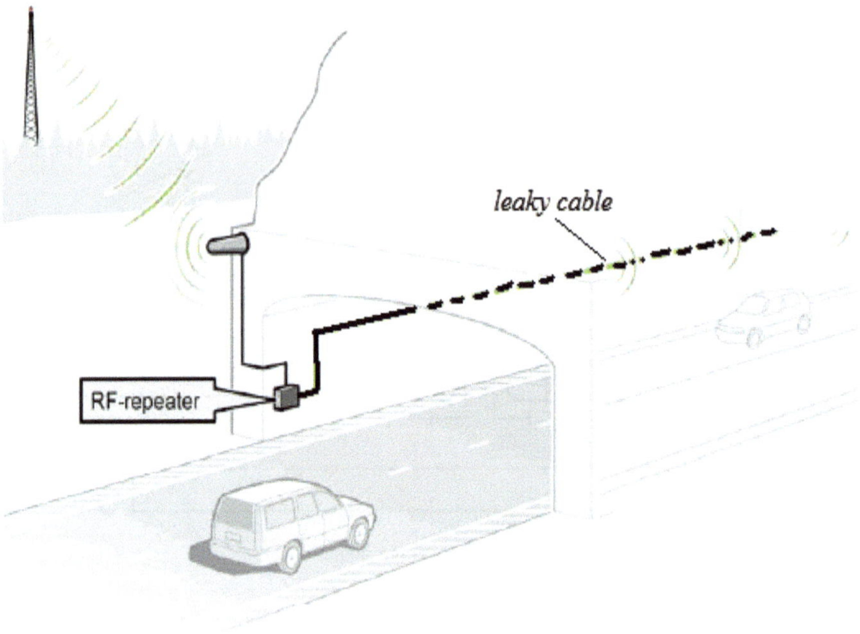

Fig. 1.14 In-tunnel indoor donor coverage extension using leaky cabling as service antenna

1.5 Traffic Considerations

Due to data service requests, in most of the cases, high-throughput connections are demanded from indoor users [3, 6] where uplink is always the weak hop. Outdoor connections, although they provide satisfactory coverage in large distances, do not provide adequate signal-to-noise ratio resulting into low throughput. When hot spot traffic buildings are covered from macro cells as an outdoor-to-indoor scenario, they provide in general a deteriorated service. As a result, indoor planning becomes a good practice for operators to increase revenue and capture high traffic. Additional data traffic generated from indoor users should beabsorbed in any case; nevertheless, having a well-planned indoor coverage means at the end of the day more traffic and longer session duration from the users inside buildings.

1.5.1 Reasoning for Indoor Coverage: Traffic Capacity

Usually, when planning indoor coverage for high traffic absorption [2], a common approach is to place many indoor cells for buildings and hot spot areas together with small outdoor micro cells and large umbrella macro cells. This is really satisfactory for GSM/general packet radio service (GPRS) for data traffic; however, it might not be a good solution for HSPA/LTE networks [3]. There are many reasoning behind this statement.

The important basic reason is the overloading due to soft handovers. In general, high data service indoor users are usually serviced by more than one cell (multilayer cell design with outdoor-to-indoor and indoor coverage). Since all these cells overlap, part of the absorbed traffic will be part of soft handover areas, resulting into high resource consumption. Specifically for WCDMA/HSPA resource means uplink interference, downlink power, and baseband NodeB capacity. The larger the overlapping percent, the worst is the resource overloading, thus offering a big challenge for planners.

Other important reason is the Rayleigh fading effect [7] due to multipath environment. Such radio environments are known as dispersive channels where radio channel characteristics depend on the time difference of the different components of the received signal. Indeed, in most mobile communications, due to location obstacles, usually there is no direct path from the eNodeB antenna and the user equipment terminal. Propagation is provided through many rays (paths); each path follows many reflections before reaching the receiver antenna. The received signal consequently is the sum of many components contributing to destructive or constructive interference. Each path is characterized by a phase shift due to distance differences ($\Delta\varphi = 2\pi\,\Delta x/\lambda + \omega t$) and a propagation delay ($\Delta t = n\,\Delta x/\upsilon$). This is called time dispersion resulting into signal degradation. Delay spread is the root mean square of each path with power P_τ, in other words a weighted average of the power and relative delay of the multipath components, and provides a measure of time dispersion:

$$\sigma_{\text{rms}} = \sqrt{\frac{\sum \tau^2 P_\tau}{\sum P_\tau} - \left(\frac{\sum \tau P_\tau}{\sum P_\tau}\right)} \tag{1.1}$$

Keep in mind that each path's power component P_τ is affected in a different way throughout transmission in the radio channel. To measure such an effect, the coherent bandwidth is defined. Coherence bandwidth defines how frequency-selective is the radio channel, or in other words how much the channel frequency response depends on frequency band. On average, the coherence bandwidth is equal to the inverse of the delay spread, thus radio designers should always confirm that delay spread < symbol time or coherence bandwidth > transmitted bandwidth. If this is the case, then always whole transmitted bandwidth fades equally, or in other words, the correlation between all frequencies in the transmitted bandwidth is high. This is called flat-fading transmission conditions as shown in Fig. 1.15.

For indoor design, typically radio channels are known as wide-sense stationary uncorrelated scattering—WSSUS. Such channels have the characteristics that, due to terminal mobility, radio channel properties depend on time and change locally within short periods of time. Long-term statistics of the channel, however, will typically remain the same on average (delay spread measures, coherence bandwidth characteristics) as shown in Fig. 1.16. A good rule of thumb is that Rayleigh fading typically occurs when a large number (> 10) of uncorrelated (out of phase) signal components are added in typical urban environment. Taking into consideration also terminal mobility received signal strength may vary as much as 40 dB (10,000 times) because of Rayleigh fading.

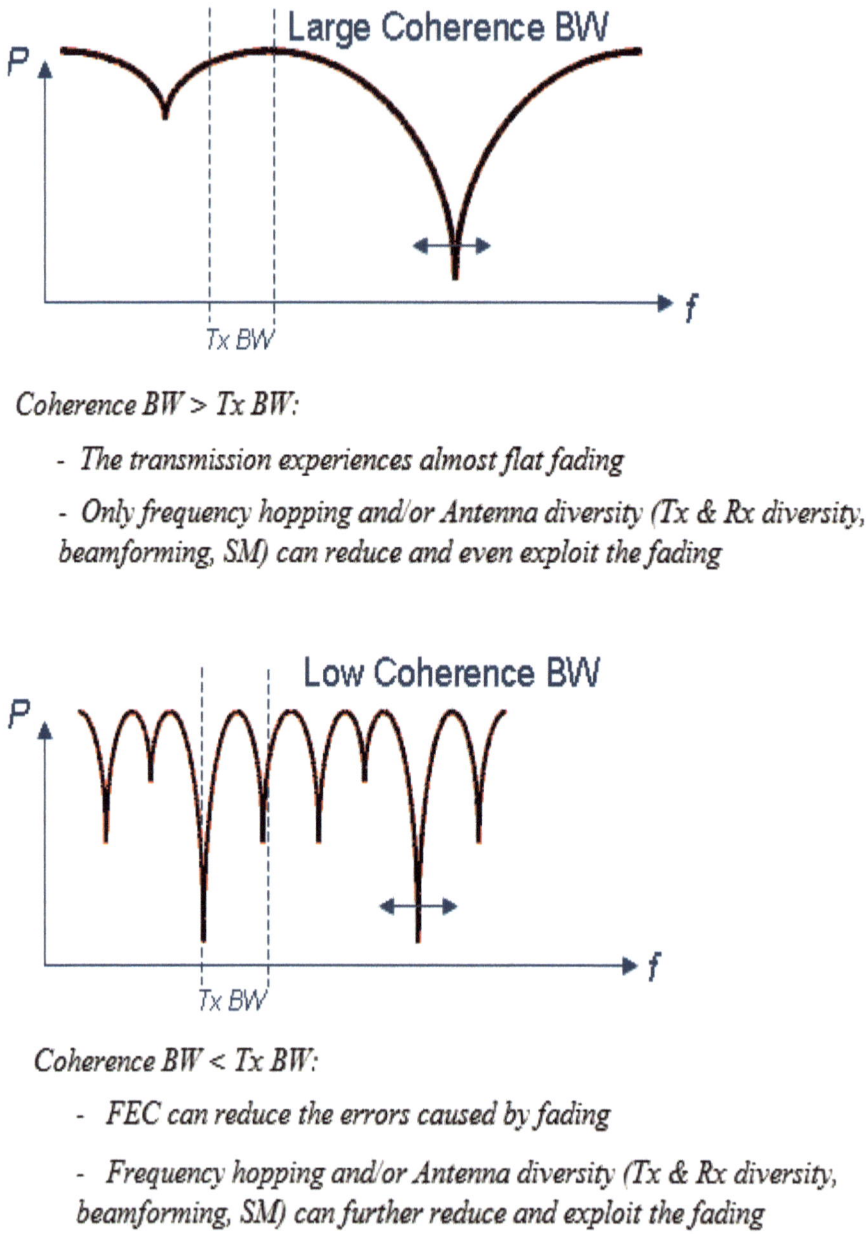

Fig. 1.15 Coherent bandwidth and delay spread conditions

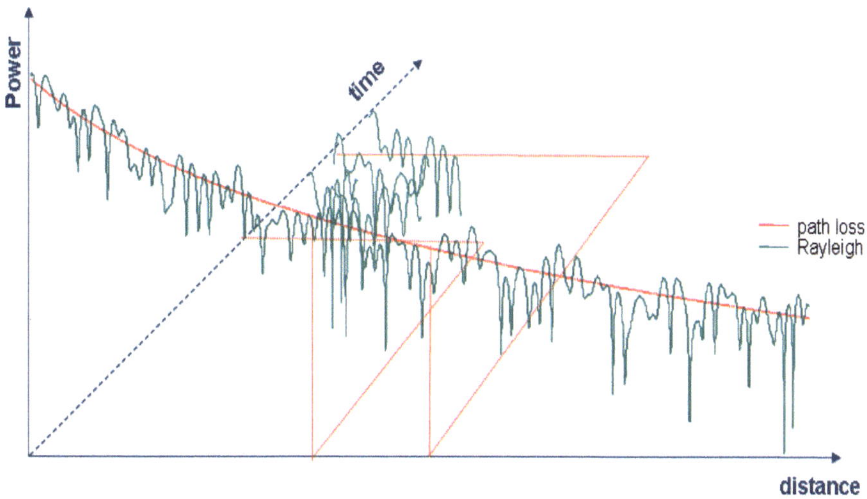

Fig. 1.16 Time varying radio channels

Another big problem is the so-called near–far problem in WCDMA/HSPA due to extensive power consumption [7]. Fading conditions affect the downlink transmission severely. In a dense urban radio environment, where indoor and outdoor-to-indoor coverage conditions hold, RF signal transmitted by a macro cell relies on many reflections when covering indoor users. Consequently, indoor users demand a large amount of downlink power; due to inner loop power control, SINR target is upgraded resulting into more downlink power for all other users in the cell. On the uplink, because of the multipath, more power is required from user equipment resulting into higher interference to serving cell and overshooting to neighbor cells. As a consequence, overall cell capacity is negatively affected and reduced.

Another sever problem for indoor users, served by outdoor-to-indoor eNodeBs, is the signal strength loss between transmitter and receiver [3]. This loss is considered to be high due to additional factors such as indoor signal penetration, internal separation walls, and furniture obstacles. If indoor coverage is not dominant inside the buildings, especially in dense urban areas, outdoor-to-indoor is the most likely connection. Penetration loss depends on many factors including kind of material, thickness, and used frequency.

At the end of the day, indoor users will have a relative high link loss and demand a high power resource both on the downlink and the uplink. These demanding users also cause other users to utilize more power, both on uplink and downlink increasing the resource capacity usage and decreasing the traffic availability and capacity [7]. Cells become faster and easier congested; it is a common sense that, especially in hot spot dense urban environments, overall required capacity (downlink power, uplink interference) could not be adequately fulfilled by the available outdoor resources.

Final problem that we will discuss is the spatial domain of the indoor radio channel. Indoor radio channel can be described in terms of two effects:

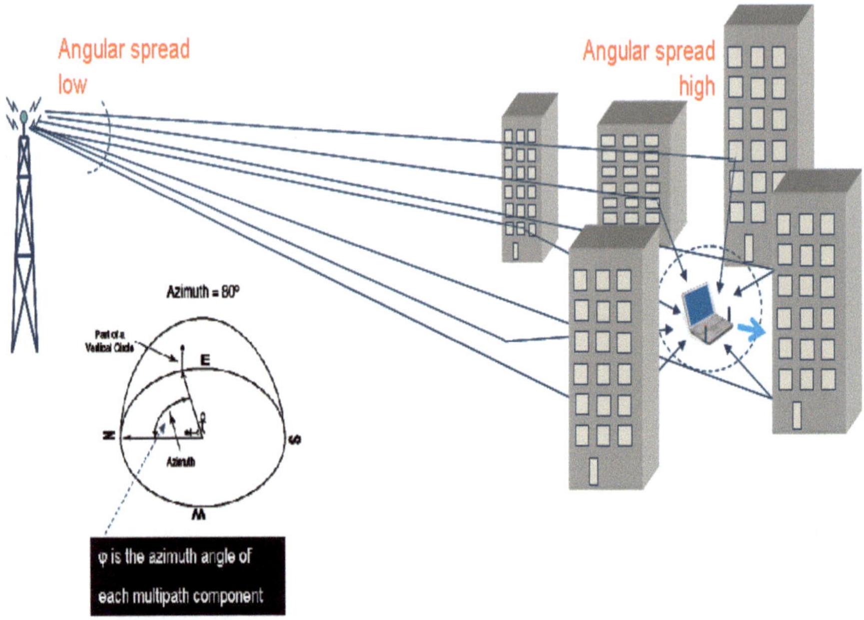

Fig. 1.17 Angular spread and radio channel characteristics

- Angular spread of direction of arrival (DOA).
- Direction of departure (DOD).

Angular spread is a weighted average of the DOA or DOD of the radio channel.

$$AS_{\mathrm{rms}} = \sqrt{\frac{\sum \phi^2 P_\tau}{\sum P_\tau} - \left(\frac{\sum \phi P_\tau}{\sum P_\tau}\right)} \tag{1.2}$$

Angular spread can be quite low at the base station (especially if it is mounted on a roof top) whereas user equipment will typically have a larger angular spread due to reflections often generated from objects surrounding the user equipment in almost all directions (internal separating walls, room objects, external building conditions) as shown in Fig. 1.17.

The major problem is that at high angular spread, conditions on user equipment receiving antenna elements become more uncorrelated (different phases) given that they are separated in the same dimension as the angular spread. This results into a degraded signal quality thus errors (bit error rate (BER)) is expected to increase resulting into low scheduling grants from MAC layer and lower throughput. The only possible solution might be beam forming on antenna.

Moreover, when network expansions or network reconfigurations [7] are performed, outdoor offloading is also suggested by indoor planning. In expansions, the most common practice is to decrease the inert-site distance by deploying more sites per planning area. Although this is a good technique for GSM networks, it is not at all recommended for WCDMA/HSPA since there is no mechanism to decrease expected interference and overshooting, especially in congested dense urban areas. Closer site distances result into higher uplink and downlink interference and inevitably into capacity congestions.

All aforementioned problems are related in dense urban environments with outdoor-to-indoor coverage. The implications to traffic capacity are obvious. The only suggestion is to provide a very well-planned indoor coverage, to offload traffic capacity from outdoor planning, and optimize both outdoor and indoor traffic capacity conditions. When indoor planning is deployed in these high-dense congested areas, the expected traffic from these indoor environments is absorbed by dedicated indoor cells. Outdoor cell layer resources are offloaded and saved for outdoor users improving and optimizing outdoor performance; outdoor resources could be used by additional users.

1.5.2 Traffic Enhancements: High Rate Services

Most of the expected high-speed data traffic is created from HSxPA or LTE. Since MAC scheduler [6] is directly involved and scheduling grant decisions are directly related to offered C/I radio conditions, a well indoor-planned environment would effectively improve the throughput of data services. In outdoor planning, the coverage region for high-throughput conditions is few hundred meters around the antenna; however, this is again dependent on the line of site probability, the geographical landscape, and the physical/technical obstacles between transmitter and receiver [8]. Reconsidering the MAC scheduler functionality in very good radio conditions, obtained when C/I > 20 dB, the channel coding rate is low and the allocated modulated scheme is rather 16-QAM to 64-QAM providing a very good bandwidth utilization of four or six bits per modulation symbol, respectively. In such conditions, the expected throughput could reach high numbers; especially for LTE. Figure 1.18 provides a measure of uplink MAC and physical layer throughputs per Scheduled Block, based on 3GPP standards [6] realized on MAC layer Link Adaptation function.

When outdoor-to-indoor coverage is the dominant case, high-throughput indoor user performance is not easily achieved mostly because of propagation conditions. Expected path loss of outer walls toward indoor user is considered always a big issue, especially when dense concrete buildings are between eNodeB antenna and user equipments on dense urban areas or old city propagation environment includes very thick brick walls as shown in Fig. 1.19. Considering LTE network, additional expected problem is the intercell interference when full 20 MHz bandwidth is used with a frequency plan of 1/1 factor [8]. Considering HSPA network additional issues are caused by uplink neighbor cell pilot pollution and the intercell interference or overshooting, downlink intracell interference due to scrambling codes andmultipath propagation phenomena that are expected to be severe in the dense urban environment.

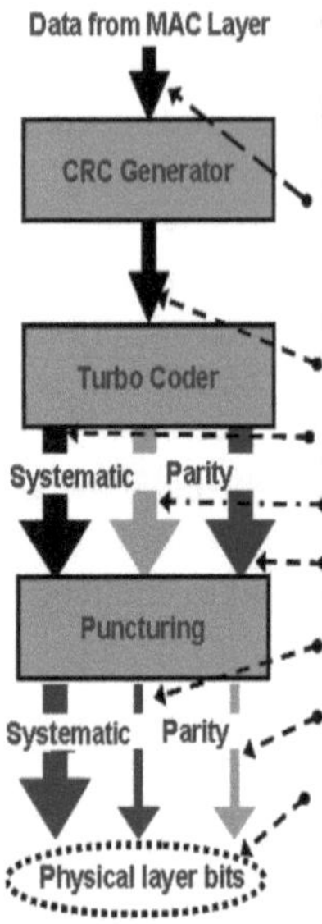

Type of Modulation	QPSK	16 QAM	64 QAM
Rate at MAC Layer [kbps]	96	432	824
Payload bits	96	432	824
CRC Size (bits)	24	24	24
Input to Turbo Coder	120	456	848
Systematic Bits (Input + 4)	124	460	852
Parity #1 Bits (Input + 4)	124	460	852
Parity #2 Bits (Input + 4)	124	460	852
Punctured Parity #1 Bits	82	58	6
Punctured Parity #2 Bits	82	58	6
Physical layer bits	288	576	864
Coding Rate [%]	33	75	95

Fig. 1.18 Uplink MAC layer Scheduled Block throughput for LTE good radio link conditions

Fig. 1.19 Nonline of site outdoor-to-indoor propagation conditions in dense urban environments

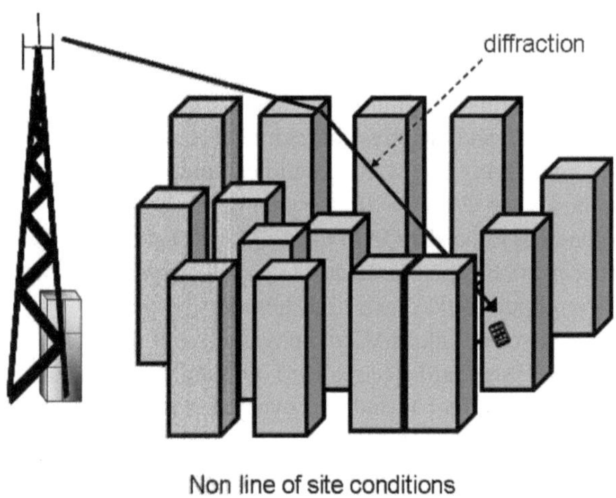

A proposed solution in such cases is again dedicated indoor coverage to enhance considerably the expected C/I ratio [2]. If this is the case, the expected modulation scheme will be quite often 64-QAM with low coding rate thus maximum throughput as shown in Fig. 1.17. If indoor cells are always the dominant case, considering indoor propagation environment protected from outdoor high penetration losses, the high C/I ratio is almost always guaranteed. If indoor design is carefully implemented and tested before turn on key phase, the serving carrier signal strength C is considered to be enhanced (very low path losses from indoor), transmitter and receiver distance is considered really small, line of site might be the case. Also, due to small distance between transmitter and receiver, lower transmission power is required contributing to lower indoor-to-outdoor intercell interference. Another important issue is the ability to replan the outdoor coverage. Indeed, when indoor coverage is really well planned, there is no need for outdoor-to-indoor services; outdoor cell coverage could be reconsidered only for outdoor users by reducing transmitted downlink power in the absence of high penetration building losses on link budget calculations. Also, tilting could be reconsidered depending on case. As a result, less interference is expected on outdoor environment improving outdoor traffic capacity [8].

The only disadvantage expected is the extremely high number of indoor reflections specially when there is no line of site, contributing to higher average Rayleigh fading conditions.

1.5.3 Traffic Enhancements: High-Speed Moving Users

Specifically for LTE users, high-speed conditions (high-speed train users, highway users) poses problems to expected throughput due to Doppler shifts. Recalling physics Doppler shift is equal to the speed divided by the wavelength of the carrier signal when the terminal is moving straight toward the transmitter, or when moving in a certain angle α toward the transmitter, the following equation is valid:

$$f_d = \frac{v}{\lambda} \cdot \cos(\alpha) \tag{1.3}$$

A good measure of channel degradation due to Doppler shifts is the channel time coherence, describing how slowly radio channel changes. The inverse of the average Doppler spread is equal to the coherence time of the radio channel:

$$T_C = \frac{1}{f_d} = \frac{\lambda}{v \cdot \cos(\alpha)} \tag{1.4}$$

Consequently, when high coherence time is measured (low-speed moving terminals) in a channel radio, channel conditions do not change rapidly. In other words, radio planners always have to guarantee that transmitted symbol period should typically be shorter than the coherence time. Otherwise, the symbols become distorted on receiver. From planning conditions there is a contradiction of desired planned levels. Based on radio channel characteristics, symbol time should be long enough in order to experience flat-fading conditions, but on the same time, symbol time should not exceed the coherence time of the channel as shown in Fig. 1.20. This is always guaranteed from 3GPP standards and this is the reason for the standardized OFDM parameters for LTE networks. From Fig. 1.19, time coherency is always guaranteed due to low-moving terminals inside indoor environments, however, the flat-fading conditions are not always the case and depend on indoor environment (obstacles, number of separation walls, and interfloor coverage).

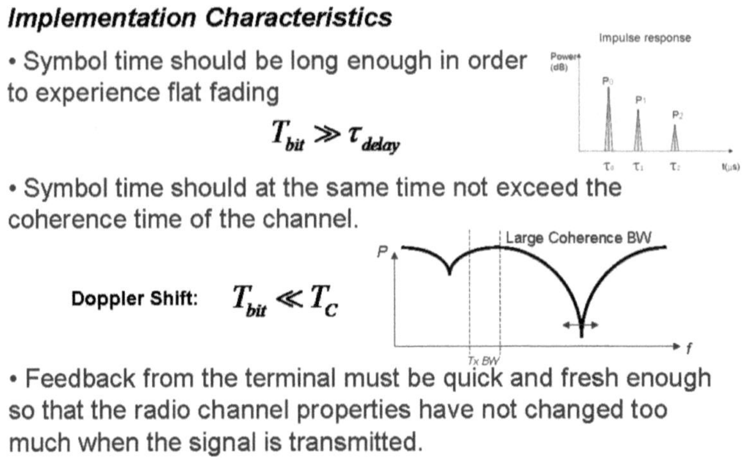

Fig. 1.20 Radio channel conditions to be fulfilled on indoor planning

Indoor radio planners should always be very careful when planning near highways [1, 7, 9]. Indoor cells should always be restricted into the indoor environment and never absorb traffic from nearby highways. Otherwise, due to low-transmission signal strength and high Doppler shifts, the expected service will result in low C/I ratio and low throughput. Moreover, the expected number of handovers will increase due to small indoor coverage range resulting in degradation of service performance (increased handover drops).

References

1. Louvros, S., & Kougias, I. (2010). *Cellular network GSM*. Athens: New Technologies Publication. Greek edition.
2. Tolstrup, M. (2008). *Indoor radio planning—A practical guide for GSM/DCS/UMTS/HSPA*. Chichester: Wiley.
3. Louvros, S., Angelis, K., & Baltagiannis, A. (2011). *LTE cell coverage planning algorithm optimizing user cell throughput*. Proceedings of 11th IEEE International Conference on Telecommunications (ConTEL 2011), pp. 51–58.
4. 3GPP (2009). TR 25.913 Feasibility Study of Evolved UTRA and UTRAN Rel-9.
5. Saunders, S. R., & Aragon-Zavala, A. (2007). *Antennas and propagation for wireless communication systems*. Chichester: Wiley.
6. 3GPP TS 36.321 (2008). Evolved Universal Terrestrial Radio Access (E-UTRA); Medium Access Control (MAC) protocol specification Release 8, V8.1.0.
7. Clint Smith, P. E., & Cervelis, C. (2007). *Cellular system design & optimzation*. New York: McGraw-Hill.
8. Louvros, S., Iossifides, A. C., Aggelis, K., Baltagiannis, A., & Economou, G. (2012). *A semi-analytical macroscopic MAC layer model for LTE uplink*. Proceedings of 5th IFIP International Conference on New Technologies, Mobility and Security.
9. Bikos, A., & Sklavos, N. (2013, March/April). LTE/SAE security issues on 4G wireless networks. *IEEE Security and Privacy, 11*(2), 55–62.

Chapter 2
VLC Technology for Indoor LTE Planning

Spiros Louvros and David Fuschelberger

Abstract Long-term evolution (LTE) indoor coverage, owing to its importance, is becoming very important for cellular operators lately. In international literature, there is a lot of research regarding visible light communication (VLC), especially indoor, to improve the expected throughput. To meet user expectations on operator's indoor high-quality services and to provide adequate capacity availability, special issues have to be studied. Indoor users expect faster Internet with less interference and healthy environment.

2.1 Introduction

Optical communications have been used in various forms for thousands of years. Since ancient years to Bell's Photophone introduced in 1880, people were using methods such as smoke or light signals in order to optically communicate over long distances. Finally, a long time later, after the invention of light amplification by stimulated emission of radiation (LASER) sources and light-emitting diodes (LEDs) in the 1960s, optical communications quickly revolutionized and spread around the world.

Today's optical communication technologies may be categorized into two groups: fiber optics and free space optics (FSO). The main difference between these two categories of optical communication technologies is the medium that each of them uses in order to propagate the data. Fiber optics is based on the optical fiber cable, which it uses as a medium, whereas FSO achieves the transmission over the air. They both use either lasers or LEDs as their sources, usually at wavelengths such as infrared (IR), visible, or ultraviolet (UV). Both technologies have advantages and disadvantages. Fiber optics, for instance, can achieve much higher data rates than FSO, whereas FSO on the other hand is capable of achieving data transmission

S. Louvros (✉) · D. Fuschelberger
Department of Telecommunication Systems and Networks (TESYD),
Technological Educational Institute of Messolonghi, Messolonghi, Greece
e-mail: splouvros@gmail.com

D. Fuschelberger
e-mail: d.fuschelberger@gmail.com

N. Sklavos et al. (eds.), *System-Level Design Methodologies for Telecommunication*,
DOI 10.1007/978-3-319-00663-5_2,

to places where either physical connection would be impossible to be deployed or absence of electromagnetic (EM) radiation is important or requested. Engineers around the world dealing with FSO fortunately tried to use LEDs at the visible spectrum as information sources anyway.

Since physics provided us with the information that common incandescent light bulbs or even the more efficient fluorescent lamps would hardly be appropriate for that purpose, because of the poor data transmission characteristics of the photons creating their illumination, the only way of using them as a source for FSO was to force them to blink, in order to decode every instance of changing light condition as a bit of information. Apparently, these kind of regular light bulbs could not be turned on and off more than a few times per second and no longer than a few seconds before they burned out. Therefore, this method would have had huge disadvantages, mainly because of data rate limitation and, furthermore, because of the fact that a blinking light bulb would have been a bad solution for communication in the first place. Fortunately, engineers focused on visible spectrum LEDs, whose light could be on–off modulated and used as a data source. One of the main restrictions in on–off modulation is the incapability of fast switching of solid-state devices, such as PN LEDs, owing to the rising time limitation. This limitation goes far inside the operating principles of LED and, to be more precise, back into quantum physics, where electrons experience a certain reaction time offset (inertia characteristics) in the presence of alternative voltages. Thus, although LED could be switched on/off several thousand times per second, this switching time is not enough to achieve thousands of megabits per second bit rate. To overcome such problems, engineers have proposed implementation of more than one communication channel at the same time. This solution could be implemented either on the optical domain or on the electrical driving circuit. On the optical domain, the solution is using different colors, since one channel's wavelength would never interfere with the wavelengths of the others. On the electrical driving circuit, this could be achieved by using either spread spectrum techniques (optical code division multiple access (OCDMA)) or multicarrier (orthogonal frequency-division multiplexing (OFDM)). Nevertheless, still the problem was the limitation in high data rates or the rise of implementation budget, in contrast to the high rates of IR sources, which were already achievable. Nowadays, LED industry has taken huge steps in developing more powerful and qualitative LED chips. Nowadays, white-colored high-power LED modules, blue LED chips combined with a yellow phosphor, are being used for indoor and outdoor illumination. Their characteristics regarding power consumption and lifetime are by far better than those of commonly known incandescent light bulbs or even fluorescent lamps. The only disadvantage of such LED modules still lies in their price and general implementation costs, but this fact is believed to rapidly change in the next few years since the market already shows a tremendous interest in using them for various purposes (e.g., TVs and displays).

This was the breaking point where a new subcategory of FSO, besides the already existing IR technology, was born: visible light communication (VLC) or also well known as wireless light communication (Wi-Li). VLC refers to data communication over a specific range of the EM spectrum, which is visible to humans. This range is measured to be approximately from 400 to 700 nm of wavelength, also known as "visible spectrum." The term "VLC" first appeared in 2003, when a small group of people at Keio University in Japan (Nakagawa Laboratory) started to experiment with LEDs and photodiodes in order to achieve communication via visible light. The Nakagawa Lab then, together with some of Japan's biggest technology firms (NEC Corp., Panasonic, and Toshiba), formed the so-called Visible Light Communication Consortium (VLCC). Later, VLCC joined forces with the corresponding IR Consortium, the Infrared Data Association (IrDA). Since then, a lot of research activities regarding VLC have been carried out around the world, with the European Framework Programme (FP) 7 OMEGA project and work done at the University of Oxford, England, being the most notable among them. Furthermore, the Institute of Electrical and Electronics Engineers (IEEE) Wireless Personal Area Network (WPAN) working group (802.15) is already working on the standardization of VLC.

What is the VLC's ultimate goal? Is it the combination of illuminating an area and providing data communication at the same time via the same technology? Furthermore, as VLC is referring to visible light, specific health-related issues have arisen against communication technologies that use radio frequency (RF), such as the broadly known and used IEEE 802.11x standards.

2.2 History of Visible Communications

One of the first implemented FSO systems was used by the French Military: *Chappe's Telegraph* system (semaphore) consisted of wooden structures mounted 5 m high every 11 km, each featuring three movable arms to create 196 different signs with word and sentence meanings as well as telescopes to observe the signs from neighboring stations in both directions. In 1 min, a single sign crossed a distance of 135 km. Lamps attached to the movable arms allowed night-time signaling.

The first experiment of VLC was exhibited by Graham Bell whose system was called Photophone. A brief description of its operation states that "Bell's Photophone made sound waves vibrate a beam of reflected sunlight." This may nowadays be understood as a simple kind of modulation. As a matter of fact, this experiment actually transmitted voice over the air long before the first radio transmission ever occurred!

During World War II, both Axis and Allies used FSO technology for certain communication, such as the German *Lichtsprechgerät 80* and the American *Infrared Telephony* device. In 1955, Zenith introduced the first wireless TV remote control *Flash-Matic Tuning*, seen in Fig. 1.4. This system used photoelectric cells in the four corners of the screen in order to control on/off, mute, and channel selection. Although, 1 year later, ultrasound technology replaced the light system, IR remote control is still common ever since. *RONJA* is a user-controlled technology project of an optical point-to-point (or point-to-multipoint) data link first deployed in Prague, Czech Republic, in 2001. The link has a 1.4 km range and a stable 10 Mbps full-duplex data rate. You can mount RONJA on your house and connect your PC or any other networking device to it. All documents for a do-it-yourself project are available for free under the GNU license.

As a conclusion of this small reference to some particular points in the history of VLC and FSO in general, one could say that this category of communication mediums has always been popular or at least considered in any way, and may be useful in many different applications in the future. Furthermore, by considering its advantages over other currently common wireless communication technologies, VLC seems to be here to stay.

2.3 Visible Light Communications: General Review

Wireless optical communication networks, when appropriately studied, developed, and optimized, could provide a reliable, high-security, interference-insensitive, and especially for elders and health-sensitive people, *biologically friendly* indoor communication and monitoring network. This network would allow the creation and expansion of seamless computing applications, telemetry, and medical sensor monitoring using large bandwidth high-frequency pulsed light instead of RFs and microwaves. VLC technology uses modulated light, emitted and received by LEDs for downlink and IR LEDs for uplink path. Both uplink and downlink could be provided sufficiently. IEEE has been working on standardization of VLC since 2009 in the context of WPANs (802.15) and recently provided a draft standard for short-range wireless optical communication using visible light, including full medium access control (MAC) and physical (PHY) layer protocols.

2.3.1 VLC: Advantages and Disadvantages

VLC is not, however, the unique existing technology for wireless optical communications. Other existing and well-appreciated radio technologies are ZigBee, Bluetooth, and WiFi. Although, nowadays radio technologies are the most dominant owing to their market penetration, especially for indoor applications, solid-state illumination technology with intensive LEDs (power LEDs) has been developed and found increasing market growth, because it *reduces significant power consumption together with expanding architectural capabilities*. It is profound that LED provides a good performance of cost versus brightness against other illumination devices [1]. LED usage (actually the whole wireless optical solution) may help in providing many services—indoor residence illumination, indoor and outdoor line-of-sight communications, area security functions, telemetry applications, and remote medical monitoring. It is well known that WiFi technology is the most dominant and worldwide respected and accepted technology among all other radio technologies [2]. In most criteria, VLC and WiFi are complementary on performance and in some aspects (power availability, Tx/Rx power, security, and data density) VLC is even superior. WiFi could be superior on range and non-line-of-sight (NLOS) radio link environments. Indeed, VLC suffers from shadowing and atmospheric absorption,

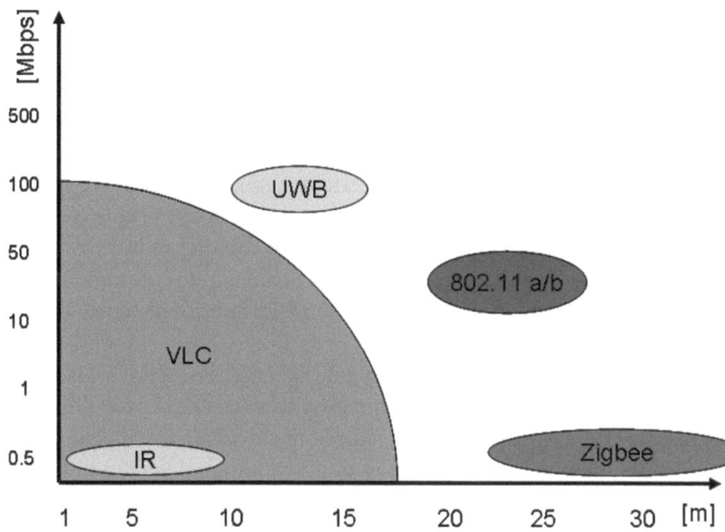

Fig. 2.1 Comparison among LTE and other existing optical wireless technologies

thus restricting its high data rate applications to short-distance communication links [1]. However, by providing appropriate indoor illumination planning and number of LED lamps in the indoor design, range will always be sufficiently small to provide enough signal strength on receiver and NLOS will never be the case. Moreover, for sufficient ranges and after proper illumination design and multiple LED arrays in the building ceiling, data rate performance is comparable to all radio technologies since distance will be eliminated owing to LED array reuse factor. To test the performance of VLC technology in several ranges, a simple PHY layer VLC prototype has been implemented, as part of undergraduate student thesis, in the Telecom Laboratories of the Department of Telecommunication Systems and Networks of Technological Educational Institute of Messolonghi. This prototype was only a simple implementation, meaning that it was not protected against visible light interference. Furthermore, neither preample electrical filter nor channel equalizer to fight back intersymbol interference (ISI) and multipath channel fadings was implemented. Under several test measurements, bit rates of 1 Mbps have been demonstrated [3]. Concluding about data rates and ranges, VLC, under appropriate illumination design and LED array distance reuse, provides superior performances over short-range radio technology competitors such as ZigBee and Bluetooth (Fig. 2.1).

The raising issue of interference of background light sources could be easily eliminated by using appropriate optical filters, a well-known technique proposed in several applications, like OMEGA FP7 project [4]. Available bandwidth, interference, and security are other issues in which VLC is considered to be superior as it can provide both partial and full solutions to a number of wireless radio environment technological problems. Such solutions include the increasingly limited availability of conventional RF bandwidths for electronic equipment, the possible communications interference with sensitive electrical equipment, and the door-to-door data security [1].

The perceived negative health consequences of existing radio technologies [1], as indicated in Fig. 2.5, when exposed to high RF and microwave levels—especially for health-sensitive human groups or buildings (schools, hospitals)—is a severe issue. From health perspective, all such applications and solutions might relax indoor space from radio emissions of contemporary telecommunication systems (WiFi, WiMax, Bluetooth, and UWB), which in certain cases are prohibitive (e.g., hospitals, airplanes, and areas inhabited by elderly people) or nondesirable (e.g., schools, university class rooms). Consequently, by replacing existing microwave-based and radio-based wireless networks, a next-generation green wireless communication network that will transform our everyday experiences and contribute to the idea of cyber green communication networks is obtained.

Moreover, in addition to all previously mentioned metrics, there is also one major advantage of optical VLC technology compared to any radio competitor. This advantage is the OFDM compatibility and superiority over VLC [5]. Indeed, OFDM enables very high data rate transmission with low computational complexity at the receiver since it is robust to multipath propagation. OFDM entirely eliminates the need for complex algorithms to cope with ISI, which typically gets worse with higher data rates. However, a standard OFDM transmitter produces a complex-valued signal. Through a simple mathematical "trick," this signal can be converted into a real-valued signal whose amplitude greatly varies in time. As a consequence, the peak-to-average-ratio (PAR) is high. This causes concerns in RF communications because of the detrimental impact on system performance due to power amplifier nonlinearities. For optical wireless communications, this effect, however, can be turned into an advantage as the high PAR signal can be exploited for intensity modulation [6]. Given that the minimum illumination for reading purposes is 400 lx, and that this already translates into a signal-to-noise ratio (SNR) greater than 30 dB, OFDM combined with higher-order modulation techniques, such as M-level quadrature amplitude modulation (QAM), results in a powerful transmission technology for incoherent visible light sources. D-Light team, University of Edinburgh, has demonstrated real-time data transmission using off-the-shelf LEDs of 130 Mbps. Finally, one last advantage of VLC compared to any other available radio technology competitor is the interference issue. Using VLC over OFDM results in high link-level data rates, making VLC a very good candidate over OFDM for indoor LTE implementation.

There is, however, one good question to be answered; what happens if multiple transmitters are deployed which together form an optical cellular network? In a recent publication [7], the area spectral efficiency (ASE) of future interference-limited wireless systems has been determined. The ASE is a measure of the maximum data rate per unit and per hertz bandwidth. It assumes a wireless network that is composed of multiple randomly deployed access points where each access point uses the same transmission resource/bandwidth. Basically, many access points means a high reuse of the same transmission resource and thus high data rate per unit area, but at a certain point this gain is outweighed by increased interference, which results in a drop in ASE. On the other hand, if there are only a few access points, this means a low resource reuse and, hence, low data rate per unit area, but also low interference.

Table 2.1 Advantages and disadvantages of VLC technology

Advantages	Disadvantages
Harmless for the human body	Atmospheric absorption
Data transmission by sockets of existing light fixtures	Shadowing/signal deterioration
Alleviation of problems associated with radio frequency (RF) communication systems	Beam dispersion
Far less energy consumption	Interference from background light sources
Increased security	
Compact integration on sensors through small dimensions	No communication if no "line of sight"
Huge number of channels available without interfering with other sources	Only discrete spectrum available as light source and sensor
Simple electronics as drive for the LEDs	Noise from interference of other sources has to be filtered
No influence to other sensitive equipment through radio waves	

Therefore, there is an optimum point for the ASE. This optimum ASE for an indoor environment is found to be 4×10^{-4} bits/s/Hz/m^2.

Implementing cellular networks and presumably LTE over VLC technology has, among other aforementioned advantages, the benefit of security and privacy. Indeed, using VLC technology, there will be no interference for indoor applications among rooms as rooms are typically separated by walls and light does not propagate through walls; an option, which does not hold in case of RF signals. If we assume a typical room of the size $4\,\text{m} \times 4\,\text{m} = 16\,\text{m}^2$, and a VLC transmitter that is capable of delivering 130 Mbps with an off-the-shelf LED lamp of 20 MHz bandwidth, as demonstrated by the D-Light team, this would result in ASE of: 130×10^6 [bits/s]/(20×10^6 [Hz] $\times$ 16 m^2) = 0.41 bits/s/Hz/m^2.

Comparing this result to the maximum 0.0004 bits/s/Hz/m^2 for state-of-the-art wireless systems, we can observe a 1,025 times higher ASE. This essentially means that VLC technology has the potential to provide wireless Gbps indoor services (over, of course, short ranges of 1–10 m) using standard off-the-shelf LEDs. This results in a massive RF spectrum relief, which frees up RF resources for the provision of better services in areas where VLC technology is difficult to use such as in remote areas. Taking this idea further and exploiting particular LED light radiation characteristics from different light sources in a room coexistence scenario, the expected ASE improvement could reach well beyond the factor of 2,000 and more.

Concluding, VLC seems to be complementary and in some aspects superior to radio technologies. WiFi might be used for wide-area coverage within a building and ZigBee for short-range communications. However, interference, unlimited bandwidth, health issues, OFDM, and security support the VLC application and data rates could be in adequate level using many VLC LED arrays for short- to medium-range indoor communications. Table 2.1 presents the major pros and cons of the VLC technology.

2.3.2 *VLC: Innovation and Standards*

Although VLC concept has its origin back to year 1880 and Alexander Graham Bell [8], the first steps of a communication system using visible light while serving illumination requirements of indoor spaces were made at the end of last century [9]. Since then, several research groups have shown great interest in modeling, analyzing, and developing prototypes in order to assess the feasibility and the performance of a VLC system. In this context, the work of Nakagawa Laboratory of Keio University was pioneering and boosted the interest in VLC [10–14]. This work led to the establishment of the VLCC Japan in 2003, which provided the first standards (JEITA CP-1221 and CP-1222) for VLC systems in 2007.

The key component of VLC systems is an LED radiating visible light, which is properly modulated for transmission information, while retaining its illumination capability. White-light LEDs are the most promising ones from illumination and communication point of view. Most research work and experiments are based on phosphorescent white LED, which consists of a blue LED chip covered with a layer of yellow phosphor. These chips are of low cost and have simpler driving, but they present a low modulation bandwidth (2–3 MHz). Proper blue-filtering before the detector allows only blue component of white light, thus increasing the bandwidth to the order of 20 MHz [15, 16] with a total achievable bit rate of 100 Mbps using discrete multitone transmission (DMT). Channel equalization techniques have also been proposed [17, 18] for further bandwidth enhancement. A good achievement of 500 Mbps over a 5 m distance has been announced in 2010 from Heinrich Hertz Institute and Siemens [19].

Since 2010, VLC has been standardized by IEEE in the context of WPAN (802.15) [20] for short-range wireless optical communication using visible light, including full MAC and PHY layer protocols. IEEE specifies three PHY layer modes: PHY I for outdoor usage, low data rate applications, on–off keying (OOK) with Manchester line coding and variable pulse position modulation (VPPM) with 4B6B line coding, data rates in the tens to hundreds of kbps, and RS outer coding; PHY II for indoor usage, moderate data rate applications, OOK with 8B10B line coding and VPPM with 4B6B line coding, data rates varying from 1.25 to 96 Mbps, and RS coding; PHY III for indoor high rate applications using color shift keying (CSK) with multiple light sources and detectors, supported data rates up to 96 Mbps, using RS coding.

Important work has also been performed by the hOME Gigabit Access (OMEGA) project [4], funded by the European Union within FP7. The project ended by March 2011 with the development of a full VLC prototype for video broadcasting and a general MAC protocol for home environments for IR, VLC, power line communications (PLC), and RF PHY layers interoperability.

2.4 LTE Implementation Over VLC: Technical Review

Regarding the general concept of VLC, as already mentioned, a primary future vision of successful implementation and usage would be the combination of LED illumination along with data access. This is what makes VLC more promising with

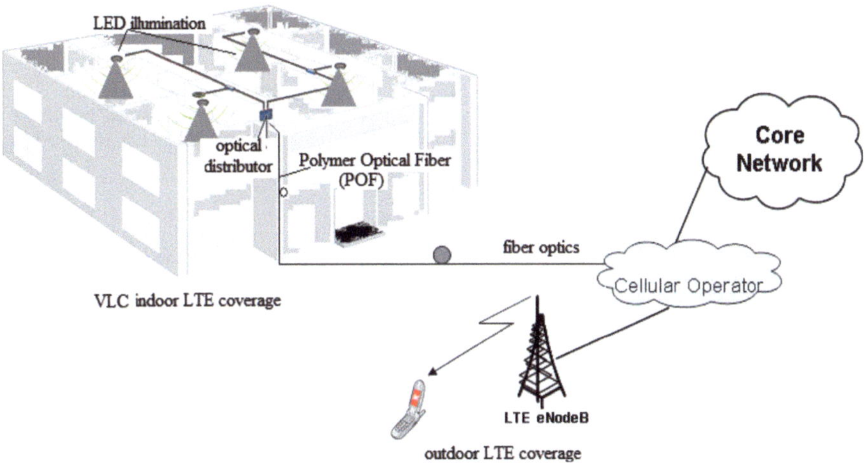

Fig. 2.2 VLC indoor LTE coverage implementation proposal

regard to its family FSO: the fact that the medium itself may be used for additional purposes in parallel.

Figure 2.2 presents the LTE indoor coverage over LTE technology proposal. VLC advantages and standards have already been presented. In such an implementation, indoor planners could make use of several existing technologies for advanced applications and purposes [21]. From outdoor cellular operator network to the indoor coverage glass, single-mode (SM) optical fibers should be used and appropriate capacity planning should be performed to provide appropriate bandwidth for optical Ethernet transmission. Inside buildings, polymer optical fibers could be used as a cheaper solution for the optical signal distribution. Optical distributors/splitters should be provided to distribute optical signal throughout building floors and rooms. Finally, LEDs with appropriate driving circuit and optical to electrical converters should also be used as the indoor distribution and illumination system.

2.4.1 Glass Optical Fiber

According to standards, LTE ideally needs Ethernet links over fiber optics. Optical fiber planning includes several stages. A wide variety of specifications shall be used at an early stage. Regarding the physical medium, system topology should be considered including cable location and/or cable routes. Existing cable protection should be carefully examined (none, building ducts, or underground ducts). Cable specifications should follow standards (fiber, moisture ingression) and number of fibers per cable shall be considered and defined. Regarding network issues, network applications and proposed topology shall be finalized together with transmission standards (SDH, SONET), available Ethernet bit rates, coding, and multiplexing.

System: 70 km span, 0.8 km between splices

Transmitter o/p power (dBm)	0	
Number of Connectors	2	In most systems only two connectors are used, one at the transmitter and one the receiver terminal.
Connector loss per connector (dB)	0.5	
Total connector loss (dB)	1	
Fibre span (km)	70	
Fibre loss (dB/Km)	0.25	
Total fibre loss (dB)	17.5	
Splice interval (Km)	0.8	Fibre is normally only available in fixed lengths up to 2 km long, so fusion splices are required, to join lengths.
Number of splices	87	
Splice loss per splice (dB)	0.04	
Total splice loss (dB)	3.46	
Dispersion penalty estimate (dB)	1.5	In buildings fibre lengths will be much shorter
Receiver sensitivity (dBm)	-30	
Power margin (dB)	6.54	Answer

Fig. 2.3 Power budget calculation including margins and losses

Specifically for the fiber itself, several parameters and characteristics have to be decided. First of all, it has to be decided whether multimode (MM) or SM shall be used, also the core size and fiber numerical aperture (NA), appropriate fiber attenuation for the link have to be determined, and budget calculations have to be done. Also, based on Ethernet-supported bit rate, fiber dispersion, including all tolerances, shall be considered. Mostly, on metropolitan networks, connectors and splitters are used quite often. For this reason, considerations on connector types, connector losses and reflections, available tolerances, mechanical or fusion splices, loss and tolerances and termination enclosures, and patch panel losses have to be calculated on link budget.

When planning is performed, always the worst case is considered, calculating also tolerances. As an example, assume the worst case transmitter output power is – 12 dBm and the worst case receiver input power needed is – 30 dBm. Then, power budget = – 12 dBm – (– 30 dBm) = 18 dB of attenuation is possible over the link before failure (nonavailability) occurs. To find maximum available fiber attenuation, we substitute from available 18 dB budget the expected loss due to connectors (connector attenuations 1 dB per connector) and splices (splice attenuation 1.5 dB per splicing) and we are left with total allowed fiber attenuation of 12.8 dB. However, this is not always enough. Indeed, we shall also add expected margins due to fiber aging (the typical operating lifetime of a communication transmission system may be as high as 20–30 years), extra future splices, extra fiber length in future operator, and maintenance repairing and extra upgrades in the bit rate or advances in multiplexing. Figure 2.3 presents an example of power budget calculation including all losses and expected margins.

In order to get a more sophisticated optical fiber planning, power penalties should be included in the analysis. Power penalty is indeed the initial calculated power increment in order to eliminate any undesired effects from expected system noise or system distortion. Most common penalties are calculated owing to fiber dispersion effects and fiber attenuation. There are two dominant fiber dispersions, time

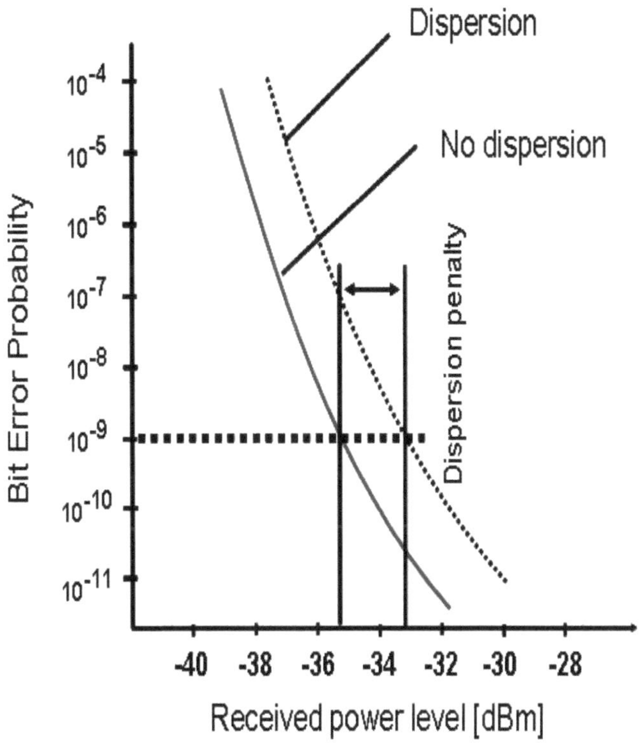

Fig. 2.4 Dispersion effect on expected quality of service BER metric and receiver sensitivity

dispersion and chromatic or material dispersion. Time dispersion is the dominant dispersion effect on MM fibers; however, when SM fiber is involved, time dispersion is almost eliminated and only the material dispersion is left. Dispersion depends mostly on supported service bit rate and on fiber material response. Whenever dispersion is present, bit error rate (BER) is expected. Figure 2.4 presents the degradation of service due to dispersion. For a specific BER performance over the transmission link, there is a difference in expected required receiver power level (also known as receiver sensitivity). Indeed, for BER $= 10^{-9}$, a typical quality of service (QoS) performance over optical links when there is no dispersion expected, the receiver sensitivity equals almost − 35.5 dBm, whereas whenever dispersion is expected, the receiver sensitivity is increased to − 33.5 dBm requesting, thus, 2 dB higher initial transmitter power or dispersion penalty!!!

Specifically for optical link planning, dispersion versus receiver sensitivity curves might not be always provided. In such a case, by reading the technical sheet of the fiber, optical plannerscould make a rough estimation of the expected dispersion penalty. There is an approximation formula to provide such estimation:

$$D_{\mathrm{chr}} = C_{\mathrm{c}} \cdot T_{\mathrm{SB}} \cdot L, \tag{2.1}$$

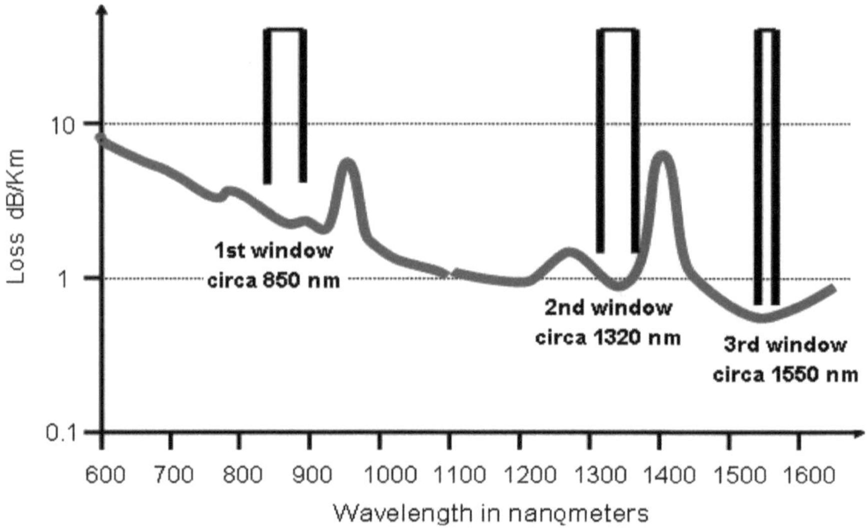

Fig. 2.5 Optical windows and fiber attenuation

where D_{chr} is the chromatic dispersion, C_c is the specific fiber chromatic coefficient in [ps/(nm · km)], T_{SB} is the transmitter (mostly solid-state laser diode) spectral bandwidth in [nm], and L is the overall fiber link length.

Losses are expected because of specific fiber material. There are specific optical windows where optical transmission shall take place with minimum losses. As a simple example on losses, Fig. 2.5 presents the optical windows. Keep in mind that such curves shall be always provided from optical fiber vendors. Most of the SM fibers on 1,550 nm spectral band, owing to industrial standards, follow ITU-T recommendation G.652 where attenuation is declared less than 0.25 dB/km and chromatic dispersion coefficient is 18 ps/(nm · km). As an example, consider a 100 km fiber link meeting G.652 standards with an oscillating cavity laser diode on 1,550 nm and spectral bandwidth of less than 0.1 nm. Then, expected chromatic dispersion would be 180 ps. Dispersion is very important to calculate because it is inversely related to expected bit rate (BR) before ISI occurs. A good approximation formula is $\mathrm{BR} = 1/4D_{chr} = 1.39$ Gbps!! To approximate power dispersion penalty, following formula provides a good estimation:

$$P_{pen}[\mathrm{dB}] = -10\log_{10}\left(1 - \frac{(\pi(\mathrm{BR}))^2 D_{chr}^2}{2}\right). \tag{2.2}$$

For optical network, depending on available supported rates, different estimations could be made. Indeed, for optical SDH (SONET) of supporting STM-1 (Fig. 2.6), STM-4 (Fig. 2.7), STM-16 (Fig. 2.8), and STM-64 (Fig. 2.9), following curves provide a graphical representation of Eq. 2.2.

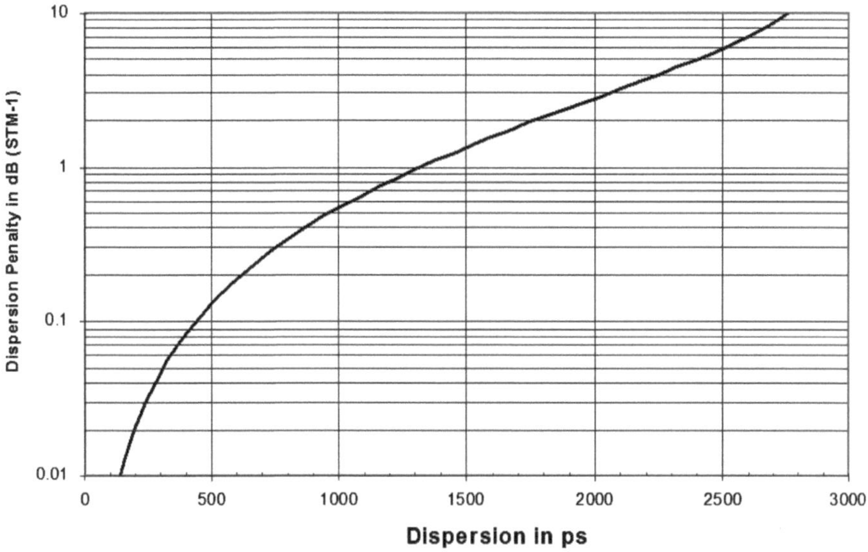

Fig. 2.6 STM-1 supported rate versus optical dispersion penalty

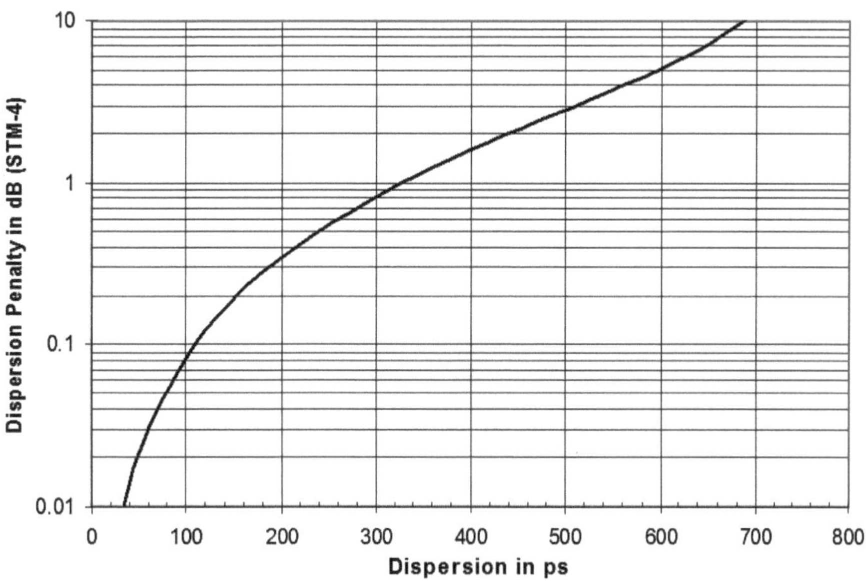

Fig. 2.7 STM-4 supported rate versus optical dispersion penalty

2.4.2 Optical Fiber Cabling

Indoor LTE radio planners over VLC, according to ISO standards, should provide appropriate cabling precautions. ISO/IEC 11801 specifies generic cabling for use

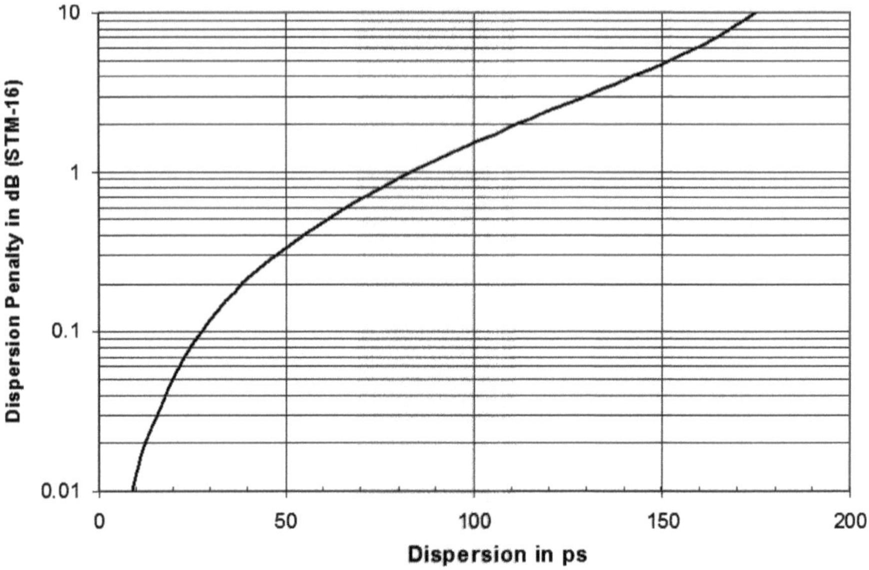

Fig. 2.8 STM-16 supported rate versus optical dispersion penalty

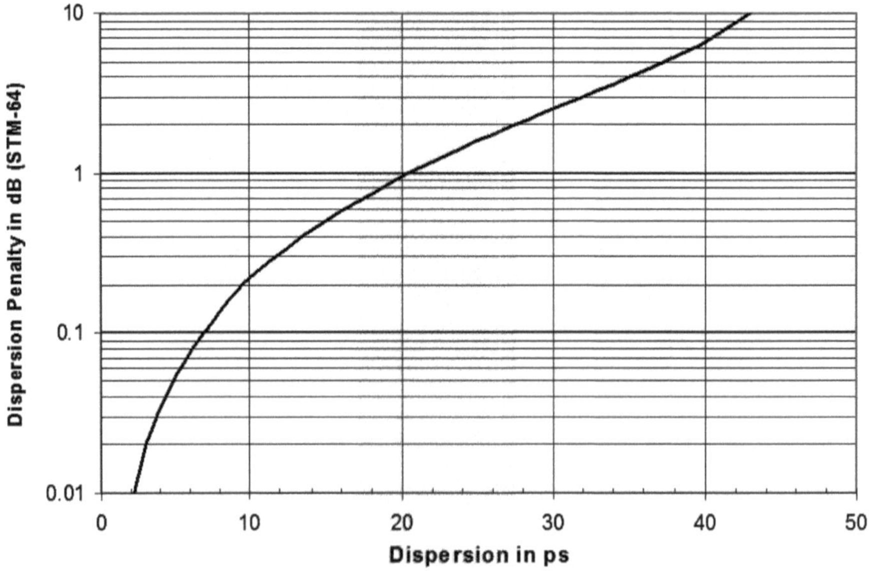

Fig. 2.9 STM-64 supported rate versus optical dispersion penalty

within premises, which may comprise single or multiple buildings on a campus. It covers balanced cabling and optical fiber cabling. ISO/IEC 11801 is optimized for premises in which the maximum distance over which telecommunications services can be distributed is 2,000 m. The principles of this International Standard may

Maximum cable attenuation dB/km				
	OM1, OM2 and OM3 Multimode		OS1 Single-mode	
Wavelength	850 nm	1300 nm	1310 nm	1550 nm
Attenuation	3.5	1.5	1.0	1.0

Fig. 2.10 Maximum cable attnuation versus different fibers

Channel Attenuation in dB				
Channel	Multimode		Single-mode	
	850nm	1300nm	1310nm	1550nm
OF-300	2.55	1.95	1.80	1.80
OF-500	3.25	2.25	2.00	2.00
OF-2000	8.50	4.50	3.50	3.50

Fig. 2.11 Fiber classes and channel attenuation

be applied to larger installations. Cabling defined by this standard supports a wide range of services including voice, data, text, image, and video. Safety (electrical safety and protection, and fire) and electromagnetic compatibility (EMC) requirements are outside the scope of this International Standard, and are covered by other standards and regulations. However, information given by this standard may be of assistance. ISO/IEC 11801 has taken into account requirements specified in application standards listed in Annexure F.

According to ISO 11801, there are three different defined and standardized fibers: OM1 fiber, 200/500 MHz · km OFL BW (in practice OM1 fibers are 62.5 μm fibers); OM2 fiber, 500/500 MHz · km OFL BW (in practice OM2 fibers are 50 μm fibers); and OM3 fiber, laser-optimized 50 mm fibers with 2,000 MHz · km EMB at 850 μm. Figure 2.10 provides optical fiber cable attenuation according to ISO 11801.

Based on the supported service and network topology, ISO 11801 defines different fiber classes. Class OF-300 supports applications to a minimum of 300 m distance, class OF-500 supports applications to a minimum of 500 m distance, and class OF-2000 supports applications to a minimum of 2,000 m distance. Figure 2.11 provides information about channel attenuation for different transmitter optical wavelengths.

2.4.3 VLC Indoor Planning Considerations

VLC uses LEDs as data sources, thus the key to success is the ever-growing LED industry and the market's turn toward them. LEDs have by far better characteristics than any other known light source in use today (except, of course, laser diodes that are

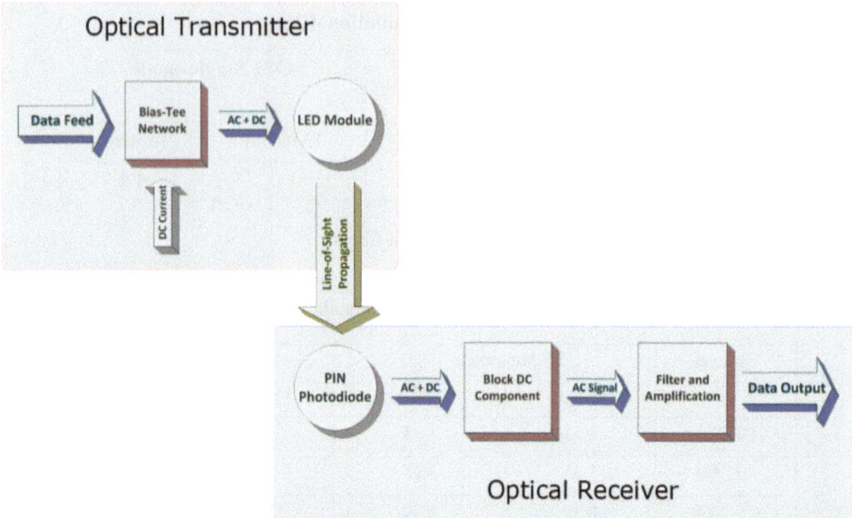

Fig. 2.12 VLC transceiver functional blocks

extremely directional, thus inappropriate for indoor illumination); energy efficiency, lifetime, and lumens output are some of the characteristics. It is estimated that in the next few years, LEDs will globally overrun the existing light sources and will dominate the world of illumination. As soon as the price of LEDs reaches a more reasonable level for consumers to prefer them to common light sources, it will be motivating for companies to start developing and implementing VLC systems, which will allow the combination of light and data in any indoor facility. Even outdoor applications may be considered as LEDs get more and more powerful. VLC could be a new milestone in local cloud access, just as 802.11x standards became a few years ago. VLC could provide adequate indoor solution for the *Last Mile* in a metropolitan area. The term "Last Mile" refers to the final connection between an internet service provider (ISP) and a customer. Initial planning goal is to achieve transmission of any signal over a wireless physical optical link by using visible light as a medium. The term "transceiver" (TRx) may be a bit misleading since it refers to a module capable of both transmitting and receiving various signals while sharing most of the required circuitry. This particular design is rather a transmitter–receiver implementation, but could eventually be extended to a TRx module by adding IR circuitry for the uplink at both the transmitter (Tx) and the receiver (Rx). VLC transmitter consists of three basic parts: the LED circuitry, the Bias-Tee network, and the transconductance amplifier (TCA) network, although the latter was not achieved to be included in the final design. On the other hand, the receiver's basic components are the photodiode and the two-stage transimpedance amplifier (TIA) network. Figure 2.12 shows the basic schematic of an indoor VLC TRx.

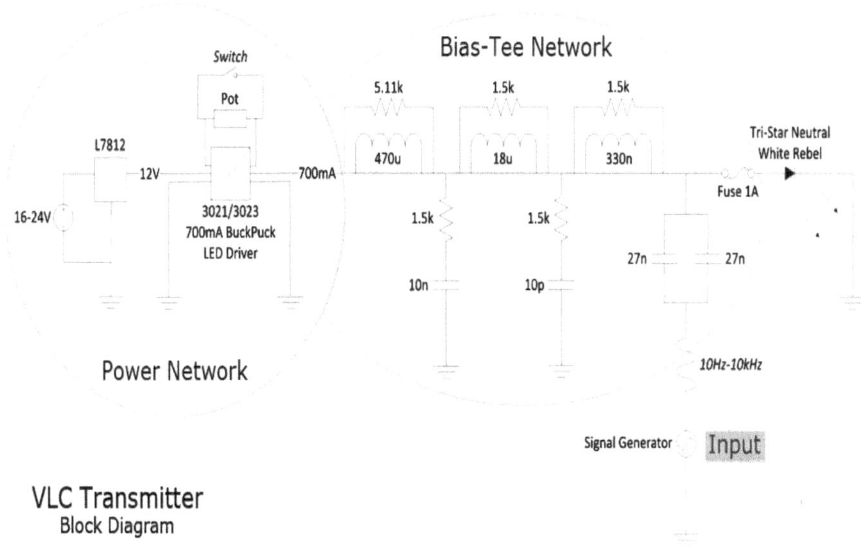

Fig. 2.13 VLC transmitter block diagram

In Technological Educational Institute of Messolonghi, a TRx circuit has been designed and tested under specific radio conditions. VLC transmitter diagram is presented in Fig. 2.13.

The power supply of the circuit had to be designed carefully, since properly regulated current is a requirement of high significance when working with LED chips. A Buck Puck driver was used in order to provide the LED module with constant DC. In order to achieve both stable consumption rate and convenient DC input values, a 12 V_{DC} voltage regulator was used to feed the LED driver (L7812), defining the input range from 16 to 24 V_{DC}. Since the forward voltage of each LED chip is 3 V at 700 mA, the tristar module has a total forward voltage of 9 V, a requirement that is met by providing the driver module with 12 V. Regarding the transmitter's power consumption, it is easily calculated being 8.4 W ($12\ V \times 0.7\ A = 8.4\ W$). The power consumption was verified by measuring the current flow of the transmitter's power supply while operating. Compared to common light sources, the ratio of lumens output and power consumption of the LED module is incredibly higher. The Bias-Tee circuit used in the architecture was borrowed from Gary W. Johnson's "Wideband Bias Tee," which was designed for a wide input range of signals [20]. Although the prototype described in this document was tested with signals of much lower frequency, a broader approach was used in order to keep potential future goals of improvement in mind. Johnson's design had a minor detail, which needed to be adjusted to the project's needs; it was designed for 250 mA. An eventual saturation of the Bias-Tee's inductors would result in worse isolation, therefore, high-quality components capable of handling up to 1 A of DC were chosen.

VLC receiver prototype was mostly based on information obtained from the work done at the University of Oxford [4], particularly, European FP7 OMEGA project deliverables. The circuit is based on receiving the light with a photodiode and turning it into an electrical signal, which in turn needs to be filtered and amplified. Photodiode is considered the most valuable component of the receiver circuit since the functionality entirely relies on its characteristics. A photodiode is a type of photodetector that converts light into either current or voltage. A common solar cell is a good example of a large-area photodiode. Apart from solar cells, two other types of photodiodes are the most common ones in use today, avalanche photodiodes (APD) and PIN photodiodes. Main differences lie, among others, in sensitivity and wavelength. APDs have better sensitivity over PIN photodiodes; owing to avalanche multiplication, they provide a built-in first-stage amplification gain when applying a high reverse bias voltage. On the other hand, PIN photodiodes are wideband and capable of achieving much higher data rates. Their physical characteristics and differences are beyond the scope of this chapter. APDs had to be rejected because of wavelength limitation; since they do not respond to the visible window, a silicon PIN photodiode was the most obvious choice. The main issue regarding photodiodes in FSO is their lack of effective active area, since they are usually coupled to a medium, such as an optical fiber, in order to focus the light beam to the very core of the diode. This is mostly the reason why arrays of photodiodes are often used in similar applications.

A very specific photodiode came as a solution to this inconvenient detail; Hamamatsu is the only optics firm, which was found to be producing and delivering photodiodes that are integrated into a concentrating lens, giving them optimal characteristics regarding directivity and active area. Specifically, the Hamamatsu Si PIN photodiode S6801/S6968 series come in a lens with a diameter of 14 mm, creating 150 mm^2 effective active area and 35° directivity, whereas their spectral response range perfectly covers visible light, making them the best choice for the prototype. The biggest benefit of using this specific component is the avoidance of a photodiode array and eventual additional components that might be required. Figure 2.14 provides the VLC receiver block circuit diagram. The receiver circuit provided by OMEGA was used in order to maintain the future target of extending this proof-of-concept prototype to a wideband application. A photodiode is usually reversely biased, in this case by 18 V_{DC}. The light captured by the photodiode is converted into electrical signals (current), which at that point are still AC/DC coupled. The DC component is eliminated by the 100 nF capacitor, whereas the remaining AC signal flows into the first stage of the TIA network where it is turned back into a voltage signal. At the second stage of the TIA network, the signal is electrically amplified and pushed to the output wire. An issue emerged regarding the power supply, with the OPA657 requiring a ± 5 V input. A Microchip TC7660 voltage converter was used in order to produce the required negative voltage, as can be seen in Fig. 3.2, but unfortunately it turned out that this component's operation was not as stable as one would like it to be. The first converter burnt just after a few times of powering the circuit. Having only one more left, it was decided to implement a back-up power supply

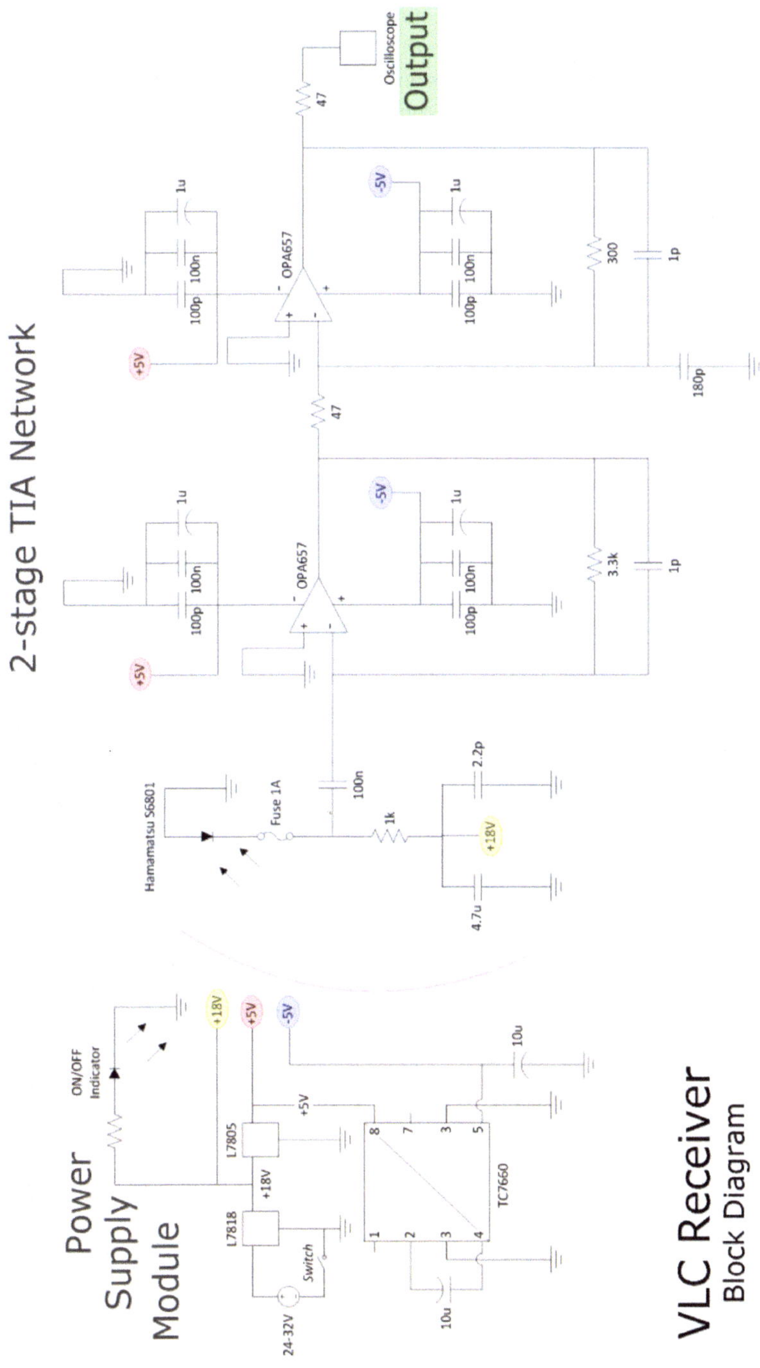

Fig. 2.14 VLC receiver block diagram

module in case the initial circuit proves totally inappropriate. The back-up power supply solution was achieved by connecting several 9 V batteries in series and taking the midpoint of first and second as the earth reference, thus creating a ± 9 V, whereas a third one provided 18 V. The ± 9 V source was then regulated down to + 5 V and − 5 V by using L7805 and L7905 voltage regulators, respectively.

References

1. Pohlmann, C. (2010). Visible light communication. http://www.slideshare.net/hossamzein/visible-light-communication. Accessed 29 June 2010.
2. VLC versus WiFi (complementary or competitors?). http://visiblelightcomm.com/vlc-versus-wifi-complementary-or-competitors/. Accessed 5 May 2011.
3. Fuschelberger, D. (2011). http://www.youtube.com/watch?v=uviWh8R-FgA. Accessed 8 Nov 2011.
4. http://www.ict-omega.eu/.
5. Differences between radio & visible light communications. http://www.purevlc.com/pureVLC_RadioTech_v1.0.pdf.
6. Afgani, M., Haas, H., Elgala, H., & Knipp, D. (2006). *Visible light communication using OFDM*. 2nd international conference on Testbeds and Research Infrastructures for the Development of Networks and Communities (TRIDENTCOM).
7. Kim, Y., Kwon, T., & Hong, D. (2010). Area spectral efficiency of shared spectrum hierarchical cell structure networks. *IEEE Transactions on Vehicular Technology, 59,* 4145–4151.
8. Kavehrad, M. (2010). Sustainable energy-efficient wireless applications using light. *IEEE Communications Magazine,* 66–73.
9. Nakagawa, M. (1999). Wireless home link. *IEICE Transactions on Communication, E82-B*(12), 1893–1896.
10. Komine, T., Tanaka, Y., Haruyama, S., & Nakagawa, M. (2001). Basic study on visible-light communication using light emitting diode illumination. Proceedings of 8th international symposium on Microwave and Optical Technology (ISMOT) (pp. 45–48). Montreal, Canada.
11. Horikawa, S., Komine, T., Haruyama, S., & Nakagawa, M. (2003). Pervasive visible light positioning system using white LED lighting. IEICE, CAS.
12. Tanaka, Y., Komine, T., Haruyama, S., & Nakagawa, M. (2003). Indoor visible light data transmission utilizing white LED light. *IEICE Transactions on Communication, E86-B*(8), 2440–2454.
13. Komine, T., & Nakagawa, M. (2003). Integrated system of white LED visible-light communication and power-line communication. *IEEE Transactions on Consumer Electronics, 49*(1), 71–79.
14. Komine, T., & Nakagawa, M. (2004). Fundamental analysis for visible-light communication system using LED lights. *IEEE Transactions on Consumer Electronics, 50*(1), 100–107.
15. Grubor, J., Gaete Jamett, O. C., Walewski, J. W., Randel, S., & Langer, K.-D. (2007). High-speed wireless indoor communication via visible light. *ITG Fachbericht, 198* (ISBN: 978–3-8007–3010-0).
16. Grubor, J., Lee, S. C. J., Langer, K.-D., Koonen, T., & Walewski, J. W. (2007). Wireless high-speed data transmission with phosphorescent white light LEDs. In proceedings of *European Conference and Exhibition of Optical Communication: Vol. 6*. PD3.6.
17. Le Minh, H., O'Brien, D., Faulkner, G., Zeng, L., Lee, K., Jung, D., & Oh, Y. (2008). High-speed visible light communications using multiple-resonant equalization. *IEEE Photonics Technology Letters, 20*(14), 1243–1245.

18. Le Minh, H., O'Brien, D., Faulkner, G., Zeng, L., Lee, K., Jung, D., Oh, Y., & Won, E. T. (2009). 100-Mb/s NRZ visible light communications using a postequalized white LED. *IEEE Photonics Technology Letters, 21*(15), 1063–1065.
19. http://www.hhi.fraunhofer.de/en/press/press-and-media/record-communication-speeds-over-ceiling-lights/.
20. Johnson, G. W. (2008). *Wideband Bias Tee*. WB9JPS.
21. Bikos, A., & Sklavos, N. (2013). LTE/SAE security issues on 4G wireless networks. *IEEE Security and Privacy, 11*(2), 55–62.

Chapter 3
Voice Over LTE (VoLTE): Service Implementation and Cell Planning Perspective

Spiros Louvros and Angelina Gkioni

Abstract This chapter provides a short introduction to the voice over long-term evolution (VoLTE). According to 3G Patent Platform (3GPP) release 8, LTE was introduced providing higher access rates and lower latency and more efficient use of radio network resources, which means lower cost per transmitted bit and voice spectral efficiency. Circuit-switched domain was excluded and all applications—services are implemented as packet-switched services. In 2009, the One Voice alternative was published by a number of communication service providers and vendors. The conclusion was that the IP multimedia subsystem (IMS)-based solution as defined by 3GPP was the best way to meet the end users' expectations for service quality, availability, and reliability when moving from existing circuit-switched telephony services to IP-based LTE services. During 2010, the "One Voice" initiative was adopted by the global system for mobile association (GSMA). That was supported by organizations, mobile service providers, vendors, and handset manufacturers. The result of this action was a GSMA VoLTE solution based on standards and supported by industry. VoLTE using IP-based service poses several restrictions on quality of service (QoS), mainly over air interface. A general model approach is also presented in this chapter, contributing to the radio network designers planning algorithms and solutions.

3.1 Introduction

The voice over long-term evolution (VoLTE) solution allows the operators to evolve from circuit-switched (CS)-based solution (mobile soft-switch solution (MSS)) in wideband code division multiple access (WCDMA) networks toward an IP multimedia subsystem (IMS)- based core network [1]. CS domain is not supported by LTE; consequently, voice service is delivered as packet-switched domain (PSD) through

S. Louvros (✉) · A. Gkioni
Department of Telecommunication Systems and Networks (TESYD),
Technological Educational Institute of Messolonghi, Hellas, Greece
e-mail: splouvros@gmail.com

A. Gkioni
e-mail: angegkio@gmail.com

N. Sklavos et al. (eds.), *System-Level Design Methodologies for Telecommunication*,
DOI 10.1007/978-3-319-00663-5_3,

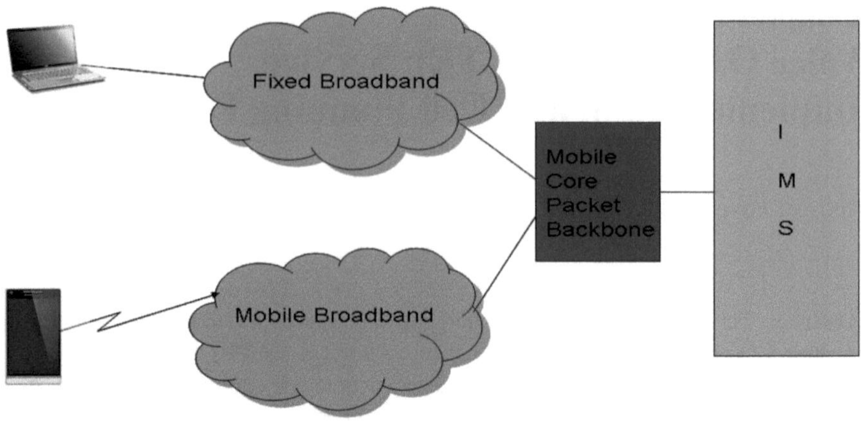

Fig. 3.1 Topology for long-term evolution (LTE) network supporting IP multimedia subsystem (IMS)

IP using the IMS-based standard. MSS and IMS use different switching nodes in the connectivity and control layer. However, mobile media gateway (M-MGW) can be used by MSS and IMS for transferring user data and signaling media payload according to 3G Patent Platform (3GPP) soft-switch-layered architecture on connectivity network layer. MSS can handle voice calls via Wi-Fi known as generic access network (VOLGA) [2] and uses IP communication between user equipment (UE) and mobile switching center server (MSC-S) being either overfixed or mobile broadband. MSS is also involved in VoLTE when there is a need to roam between operators [3].

Media gateway control function (MGCF), a standard switch node in the IMS, which communicates with the call session control function (CSCF) and controls the connections for media channels in an IMS media gateway (IMS-MGW). MGCF performs protocol conversion between ISDN user part (ISUP) and the IMS call-control protocols. The MGCF can be embedded in MSC-S whereas the equipment needed for IMS connections is located in the firmware of MGW, known as IMS-MGW, being controlled by MGCF. IMS-MGW is a component located in the IMS 3G architecture, which could terminate bearer channels from a CS network and/or media streams from a packet network. It supports media conversion, bearer control, and payload processing (e.g., using codecs, echo cancellers, or conference bridges; Fig. 3.1).

As a result, we may summarize and say that IMS offers *a standardized, future-proof network architecture* with open interfaces, guaranteeing interoperability in multivendor environments and smooth evolution with maximum reuse of the existing networks. These advantages result in a smooth and operator-decided evolutionary path from mobile soft switching (MMS) to IMS within the same Open-Core system.

3.2 VoLTE Using IMS Overview

VoLTE [4] is based on multimedia telephony (MMTel) service, a standardized IMS-based voice over Internet Protocol (VoIP) solution where evolved packet core (EPC) domain provides access mobility connectivity and the IMS domain provides the call control functionality. Packet domain network gateway (PGW) node establishes a diameter session toward policy and charging rule function (PCRF) over Gx interface. PCRF is mandatory for IMS voice sessions since bearer-level quality of service (QoS) is based on the IMS voice media session information, negotiated for each voice call dynamically. For voice call, PCRF also gets voice session information from the PCSCF and based on that information the PCRF decides the required QoS for a bearer to be established and this information is sent toward PGW.

LTE uses the concept of bearers to carry data between UE and core network and to provide QoS differentiation. QoS means that different services have different characteristics and demand different amount of data, bit rates, and transport capabilities. The new dedicated bearer will have a QoS class identifier (QCI) value according to negotiated QoS in PCRF. This QCI is connected over radio domain to scheduling priorities on eNodeB. Session management functionality in mobility management entity (MME) establishes and handles connections between UE and a packet data network (PDN). Each PDN connection is assigned at least one bearer known as default bearer; depending on the service the connection is used for, a number of dedicated bearers is also established. Default bearer is a nonguaranteed bit rate (non-GBR) bearer whereas dedicated bearer can be either a GBR or a non-GBR bearer. PGW also sends a proxy CSCF (P-CSCF) address in the PDN connectivity response that is also signaled to UE. An LTE user has at least one PDN connection, when the user is registered into the evolved packet system (EPS) network. Specifically for VoLTE users, PDN connection is always established toward a well-known IMS access point name (APN). APN will define the entry point into IMS domain, PDN. Consequently, "connectivity request" procedure is standardized to be used by UE, during international mobile subscriber identity (IMSI) attach procedure, to request the setup of a default EPS bearer to a PDN. PGW allocates an IP address (Ipv4 or Ipv6) to be used by UE. The PGW also sends a P-CSCF address in the PDN connectivity response that is signaled to the UE. When the PGW sends a request to PCRF, it contains at least the UE IP address and radio access type. During PDN connection, the PCRF may provide policy and charging control rules, for example, the QCI to apply to a specific bearer.

Initial QoS settings of default bearer are assigned by the network based on subscription data. The number of dedicated established bearers can vary throughout the lifetime of the connection. A tunnel between UE and S/P GW for all bearers (default or dedicated) is assigned. According to 3GPP PDN, connection is used to transport payload between the UE and the PDN, using one or more available EPS bearers. Information about the PDN connection is stored in the MME, according to Fig. 3.2, to be used when connections or bearers are, for example, modified or deactivated. According to standards and specifically for VoLTE solution, default bearer is used for IMS signaling and has $QCI = 5$.

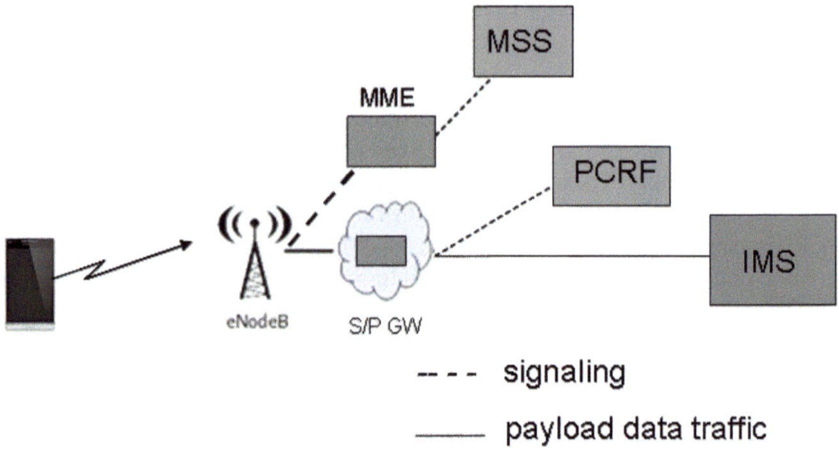

Fig. 3.2 Topology for IP multimedia subsystem (IMS) platform

3.3 IMS Multimedia Telephony (MMTel)

MMTel is an end-to-end network solution allowing operators to offer the same service over many different access types. It is a global standard based on the IMS, offering converged, fixed, and mobile real-time multimedia communication using the media capabilities such as voice, real-time video, text, file transfer, and sharing of pictures, audio, and video clips. With MMTel, users are capable to add/drop media during a session. The user can start with chat, add voice (for instance Mobile VoIP), add another caller, add video, share media and transfer files, and drop any of these without losing or having to end the session. IMS/MMTel service consists of basic communication information and optional supplementary services. MMTel is prepared to handle voice calls from any UE supporting session-initiated protocol (SIP). Since there is a standardized network-to-network interface (NNI) in MMTel, it is possible to interconnect all the multimedia features enabling operators to become world's largest multimedia community. In operator and vendor market perspectives, MMTel has been positioned as a future mass-market service for real-time multimedia communication. At launch time, around 2011–2012, initial MMTel service community was small; consequently, to fit and be compatible with previous network topologies, it will have to interwork with existing mass-market services. The 3GPP has also standardized interworking implementations between MMTel voice/video and circuit-switched video telephony, as used in other 3GPP networks such as WCDMA or global system for mobile communications (GSM)/general packet radio service (GPRS).

From network planning point of view, it is important to remember that 2G and 3G technologies evolve in parallel with 4G in existing operator networks. The consequence is that users can get their CS (voice, video) communication needs met by

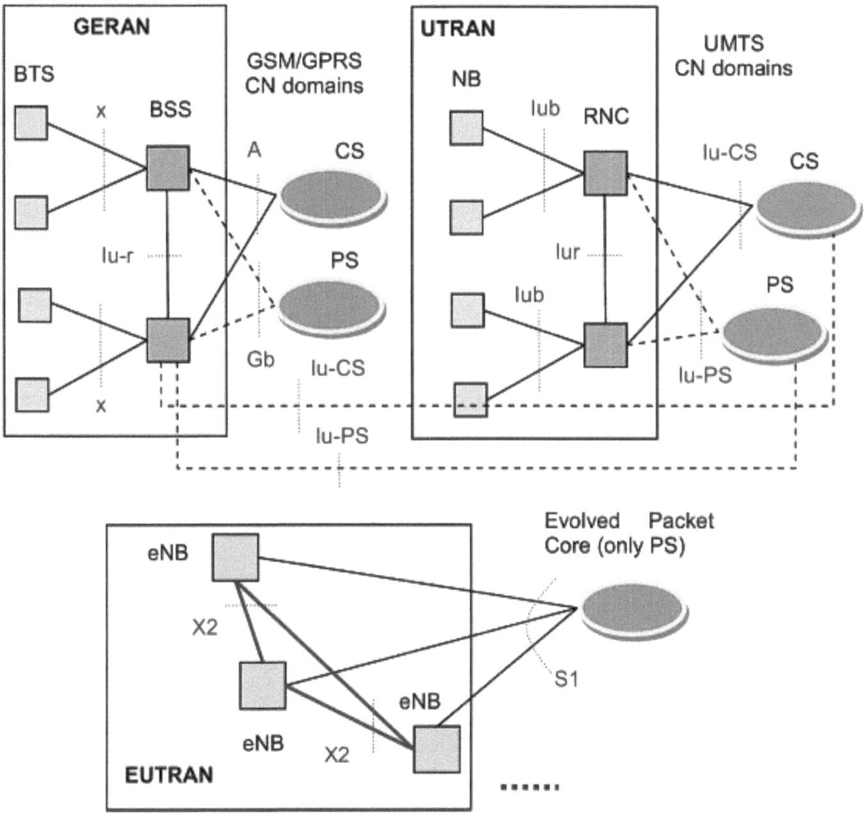

Fig. 3.3 Radio access networks (RANs) overview for cellular services

2G and 3G instead of 4G, depending on how much bandwidth they demand and how much bandwidth is available due to other existing data services (Fig. 3.3).

There are different options defining the implemented solution for a user, initially connected to LTE (VoLTE), to fallback to another technology 2G or 3G. Such options are:

- Simultaneously 2G/3G voice and 4G data (SVLTE).
- CS fallback (CSFB).
- Single radio voice call continuity (SRVCC).

3.3.1 SVLTE Solutions

During SVLTE, the handset works simultaneously in both LTE and CS mode (Fig. 3.4).

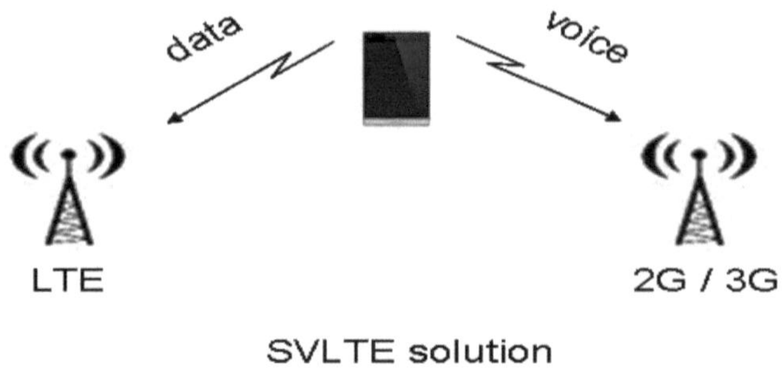

Fig. 3.4 Long-term evolution (LTE) connectivity options in 2G/3G

The LTE mode provides data services and the CS mode provides voice services. SVLTE uses different antennas for CS connections over 2G/3G and packet-switched (PS) connections over LTE. SVLTE is based on handset that will be able to support connection to both LTE and 2G/3G at the same time, and does not put any special requirements on the network. However, these handsets could be expensive and consume more power.

3.3.2 CSFB Solution

CSFB is an interim solution in 3GPP Release 8. In this approach, the LTE provides data services and when a voice call is to be initiated or received, it falls back to the CS domain. When using this solution [5], operators need to upgrade the MSC instead of deploying the IMS and, therefore, can provide services quickly. However, the disadvantage is longer call setup delay (Fig. 3.5).

There are, however, specified requirements to fall back to a CS voice service in a GSM or WCDMA network, if, of course, available in the same coverage area. CSFB requires that:

- SGs interface between MME and MSC-S is defined.
- The UE is dual-radio capable.
- UE is registered in CS domain.
- The MSC-S is updated and allows the CSFB to perform paging via LTE.

Considering LTE implementation and operator budget restrictions, it is expected that LTE will initially provide coverage to small geographical areas with high user capacity, trying to absorb high data traffic demands as much as possible. Such areas are city centers, malls, stadiums, business districts, etc. In such areas, service continuity must be provided when the subscriber is moving outside LTE's coverage area. In CSFB option, the user connected to LTE will be redirected by the network to cell reselection, from 4G to 2G (GERAN) or 3G (UTRAN) access network, to connect to the CS domain when there is a call request. The advantage of such network functionality performance is that CSFB might be used as a generic telephony fallback

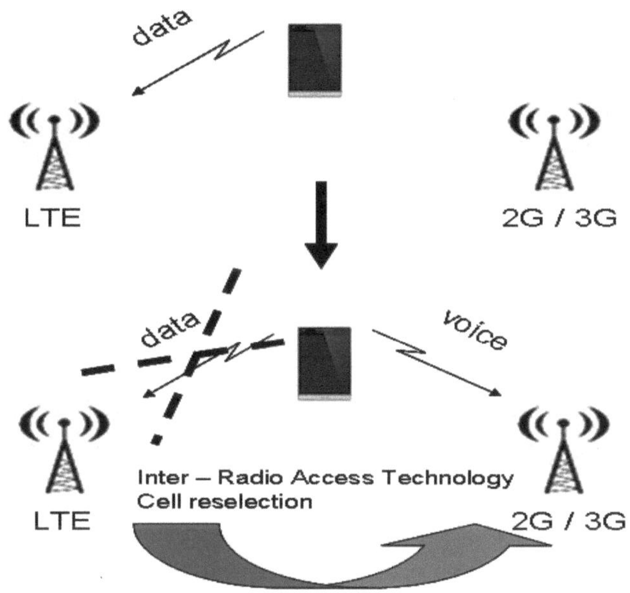

Fig. 3.5 Long-term evolution (LTE) connectivity options in 2G/3G

method, securing functionality for incoming roamers as well. The registration, location update, and paging procedures are the same for SMS and CSFB. From signaling messages and signaling needs, without explaining all details, CSFB performs the following steps:

1. Subscriber is registered in MSC, but roams in LTE.
2. Incoming call to subscriber in LTE.
3. MSC paging over SGs, S1, and enhanced Uu 3GPP (eUu) interface (Uu interface is the radio interface between the mobile and the radio access network eNodeB).
4. UE notifies MME with an extended service request.
5. MME orders eNodeB to release and inform UE that CSFB is ok.
6. UE and RAN trigger an enhanced release with redirect.
7. UE sends a location update and call setup over 2G/3G radio (Fig. 3.6).

3.3.3 VoLTE Using MMTel

In order to provide call continuity and avoid any interruptions such as interradio access technology (IRAT) cell changes or CSFB transfer, VoLTE can perform handover transfer of ongoing voice call from LTE (EPS/IMS domain) to CS domain in case native VoIP connection can no longer be maintained in LTE. This procedure is standardized as SRVCC, which means that only one radio technology applies at the same time in the UE and the services continue (handover), but is

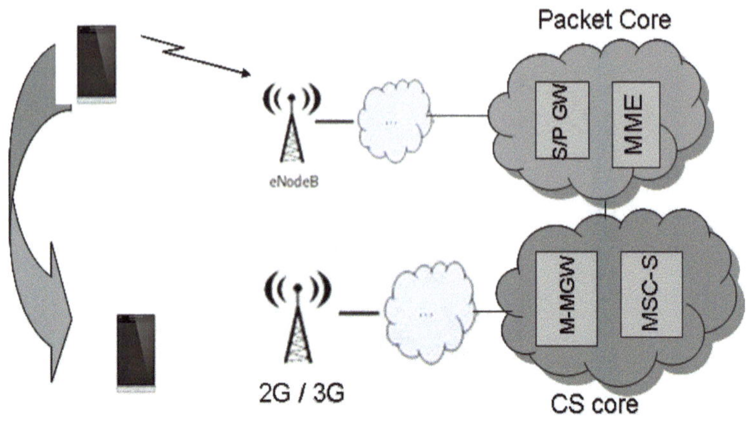

Fig. 3.6 Long-term evolution (LTE) circuit-switched fallback (CSFB) solution

also known as PS–CS access domain transfer. SRVCC technology enables handover voice communication between VoLTE and CS in 3G. Enhanced SRVCC (eSRVCC) is developed to facilitate the handover from VoLTE to CS domain (GERAN/UTRAN; Fig. 3.7).

In order to support access domain transfer, new functionalities are required. An SRVCC-enhanced MSC-S that is able to perform required procedures toward CS domain and also enhanced EPS and more specific MME and eNodeB. UEs will also have to specifically support the SRVCC function. It is important to remember that one of the main advantages of such a solution is that only Voice part of communication between UE and network is transferred from LTE to CS; other data bearers are maintained via the PS core.

Shortly describing, SRVCC-enhanced MSC-S prepares resources toward IuCS or an interface for immediate domain transfer and updates remote side of session after domain transfer. After CS resources have been committed, the SRVCC-enhanced MSC-S will establish call on behalf of UE to a specific address given by MME over Sv interface (Fig. 3.8).

When VoLTE UE makes *an initial attachment*, it indicates its voice domain preference (e.g., IMS PS voice over CS or only PS voice-capable) and SRVCC capability. The MME uses the UE-provided information, subscription information, local policy, and SRVCC capability of the network to decide if VoIP can be provided. The decision is signaled back to the UE in *attach accept message*. If VoIP can be supported, the UE initiates IMS registration in order to initiate and receive VoIP communication, otherwise UE will try CSFB to GSM or UTRAN. In order to facilitate the session transfer (SRVCC) of the voice component to the CS domain, the IMS MMTel sessions need to be anchored in the IMS. IMS asks MSC-S, via Packet Core, to take over the call. Services can still be provided from IMS, even though the subscriber is not handled by IMS. This is the purpose of IMS centralized services (ICS).

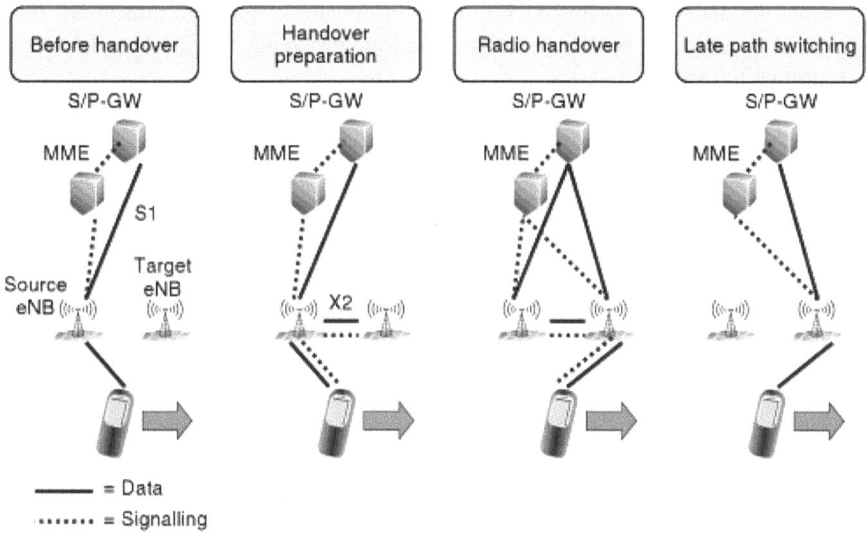

Fig. 3.7 Long-term evolution (LTE) radio access network (RAN) mobility overview

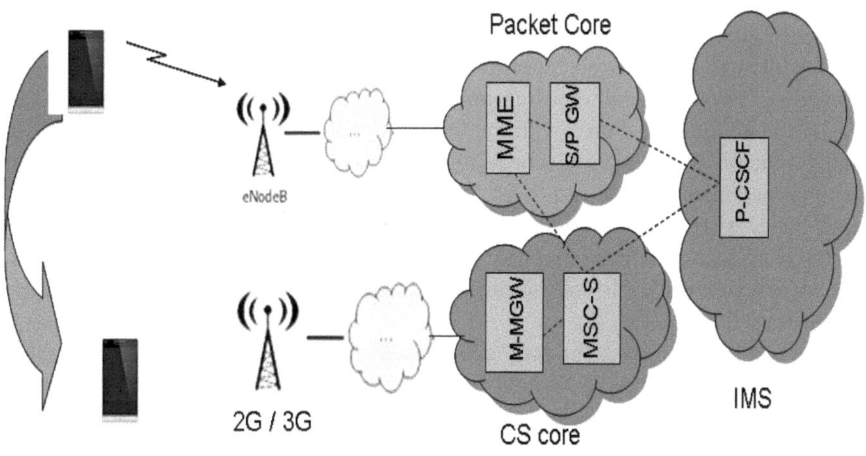

Fig. 3.8 Single radio voice call continuity (SRVCC) solution

3.4 LTE Air Interface Overview

A new generation of wireless cellular networks, called enhanced UTRAN (E-UTRAN) or LTE workgroup of 3GPP, has been evolved providing advantages to services and users over broadband wireless [6, 21]. Advantages of LTE, compared with older broadband technologies, are strongly dependant on throughput and latency requirements. As an overall enhancement, E-UTRAN should be able to support average data rates of up to 300 Mbps in the downlink and 50 Mbps in the uplink.

Considering also a 20 MHz FDD uplink and downlink spectrum allocation, it is expected to achieve 5 bps/Hz downlink spectrum efficiency and 2.5 bps/Hz uplink spectrum efficiency.

In typical LTE deployment, particularly in the eNodeB/cell dimensioning process, typical network requirements are coverage area, number of subscribers (also known in international literature as traffic load), traffic type (QoS requirements from the core network), traffic model, transmission power, and uplink/downlink cell-edge throughput. Coverage area is the first parameter that a network planner is considering in order to be able to calculate the expected neighbor cell interference (also known as intercell interference) in the serving cell. Expected number of subscribers in the serving cell is also important factor since it provides a measure of how often users are scheduled and how often the common resources are used, also contributing to intercell interference factors. Traffic type (also known as service-like VoIP) is also of important consideration since it provides a measure of the expected supported bit rate in the serving cell area of coverage. It is directly interconnected to the QoS of the supported cellular service and it is an input to the scheduler. LTE QoS complies with the 3GPP Rel 8 TS 23.203 QoS concept, providing priorities to different services. Different services are supported in 3GPP LTE and each one with a different QoS profile; however, all of them are based on IP. Traffic type and number of subscribers comprise an expected traffic load (in Erlangs), which influences, as an overall factor, the scheduling usage and the intercell interference. Transmission power is also affecting intercell interference; however, it is not of such importance for the scheduling procedure. Finally, in radio design process, cell edge should mostly be studied as this is the geographical area with the lowest signal-to-interference and noise ratio (SINR) factor, thus affecting scheduling and throughput. As a consequence, radio air interface must be able to provide both high-peak bit rates and acceptable cell-edge bit rates. However, besides the cell-edge throughput requirements, during the deployment and dimensioning process, important attention should be also given to the latency, especially for VoIP services, in order to provide an overall sufficient and satisfactory QoS.

The 3GPP LTE is based on orthogonal frequency-division multiplexing (OFDM) principle over air interface. OFDM is in principle an efficient modulation scheme (and not a multiple access scheme as it is commonly referred in international literature) where each user has been allocated part of existing bandwidth (called subband) in specific time instances. OFDM principle divides allocated frequency band into a number of narrow 15 kHz subcarriers. A minimum group of permitted subcarriers consists of 12 subcarriers of 180 kHz bandwidth. As modulation quadrature phase shift keying (QPSK), 16QAM, or 64QAM might be used depending on the channel conditions, thus representing different number of bits into OFDM symbols (QPSK 2 bits/symbol, 16QAM 4 bits/symbol, and 64QAM 6 bits/symbol). Due to time-dispersive radio channels, where intersymbol interference (ISI) is present, a cyclic prefix as a time-offset is added to the OFDM symbol duration to maintain the time orthogonality between subcarriers, resulting into an OFDM symbol duration of $1/\Delta f +$ cyclic prefix. Each OFDM symbol is known as resource element. The transmission of information, meaning what resource a scheduler could schedule

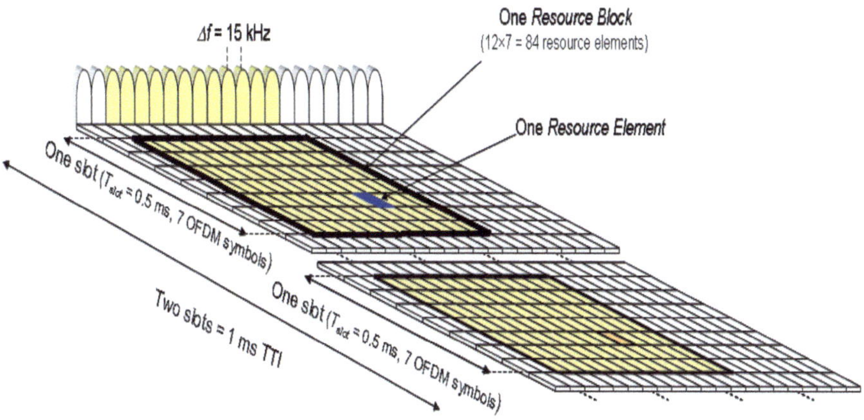

Fig. 3.9 Orthogonal frequency-division multiplexing (OFDM) principle: long-term evolution (LTE) air interface

for transmission, is known as resource block (RB). One RB is a two-dimensional resource, which has a total size of 180 kHz (12 subcarriers of 15 kHz each) in the frequency domain and seven resource elements (RE) per subcarrier of duration 0.5 ms in the time domain, thus one RB has $12 \times 7 = 84$ RE. Transmission over radio interface consists of two RBs known as 1 ms transmission time interval (TTI), as presented in Fig. 3.9. Allocated bandwidth consists of $N \times 180$ kHz, is defined per sector and it could belong to a wider bandwidth, which is allocated to the operator.

If neighboring cells are using same bandwidth or nearby spectrum, the system is expected to be strongly influenced by interference, resulting into lower capacity (Shannon limit) and into lower throughput. Frequency reuse patterns and also more clever frequency handling techniques have been recently proposed and considered in international literature since they are connected directly to the SINR factor. LTE is indeed the evolution of high-speed packet access (HSPA) cellular networks toward 4G [6]. It provides backward compatibility to GSM-WCDMA existing networks and at the same time it is functionally compatible with WiMAX networks, easing the network planners for a broadband network heterogeneous planning convergence [7].

3.5 VoIP Quality of Service

One of the first considerations for cell planners is to optimize geographical coverage and provide adequate QoS based on operator restrictions. Frequency reuse has been standardized by 3GPP [8–11] and also recently has been considered in international literature [12, 13]. MAC scheduler [20], being responsible for the

QCI	Resource Type	Priority	Packet Delay Budget	Packet Error Loss Rate	Example Services
1	GBR	2	100 ms	10^{-2}	Conversational Voice
2		4	150 ms	10^{-3}	Conversational Video (Live Streaming)
3		3	50 ms	10^{-3}	Real Time Gaming
4		5	300 ms	10^{-6}	Non-Conversational Video (Buffered Streaming)
5	Non-GBR	1	100 ms	10^{-6}	IMS Signalling
6		6	300 ms	10^{-6}	Video (Buffered Streaming), TCP-based (www, ftp, e-mail, chat, p2p file sharing, progressive video, etc.)
7		7	100 ms	10^{-3}	Voice, Video (Live Streaming) Interactive Gaming
8		8	300 ms	10^{-6}	Video (Buffered Streaming), TCP-based (www, e-mail, chat, ftp, p2p file sharing, progressive video, etc.)
9		9			

Fig. 3.10 3GPP quality of service (QoS) standards per QoS class identifier (QCI) and per service: voice over Internet Protocol (VoIP) is guaranteed bit rate (GBR) priority 2

dynamic allocation of frequency-time resources into many users, provides uplink decisions mainly based on:

- Signal-to-noise and interference ratio γ_{RB} measurements per RB.
- Required QoS received from core network (QCI) [14].
- Cell load conditions (including interference and availability on RB).
- Delay constraints.

3.5.1 Service Requirements

To provide efficient resource usage for VoIP services, LTE concept supports fast scheduling [15] considering the importance of calculating radio delay mostly due to scheduling decisions. Following Fig. 3.10 according to 3GPP standards, VoIP delay is considered to be around 100 ms.

3.5.2 Cell Planning Process for VoIP Services

Cell planning process is important to include VoIP delay constraints. Following steps should be considered:

- Path loss estimation.
  Initially, consider the operator-selected cell range as initial important coverage constraint. Mainly due to operator-determined restrictions regarding user-

estimated throughput, path loss $L_{celledge}$ at worst radio conditions (cell-edge user for outdoor planning) have to be estimated. Estimations should be based on certain defined path loss models, where a well-defined formula for 2.5 GHz LTE microcell outdoor-to-outdoor coverage is [16]:

$$L[\mathrm{d}B] = \begin{Bmatrix} 39 + 20\log_{10}(d[\mathrm{m}]), & 10\,m < d \leq 45\ \mathrm{m} \\ -39 + 67\log_{10}(d[\mathrm{m}]), & d > 45\ \mathrm{m} \end{Bmatrix} \tag{3.1}$$

- The noise floor per RB N_{RB} has to be calculated.
  Noise N_{RB} per RB is considered to be – 174 dB/Hz and for 180 kHz RB bandwidth it is calculated as – 111.44 dB [17].
- Intercell interference.
  At worst, cell conditions for outdoor planning (cell-edge user) uplink interference per RB has to be calculated. Interference is mainly considered to be intercell interference from a neighbor cell. From cell planner perspective, we do consider that it is more accurate to have an average estimation of intercell interference per RB, at a given path loss, from real-drive test measurements. Appropriate plots of absolute interference per RB versus cell-edge Path Loss L_{target} have been created from drive test according to Fig. 3.11.
- Uplink signal-to-noise and interference ratio γ at cell edge.
  Uplink γ ratio is extremely important to be estimated as it is directly related to MAC scheduler link adaptation [20], affecting the RB selection on uplink scheduling. Of course, on cell-edge conditions (worst conditions for outdoor planning) the target γ has to be always higher than the eNB receiver sensitivity S_{eNodeB}, which is defined as the minimum received power on RBS required to correctly decode uplink RB with 1×10^{-10} bit error rate (BER) [17]:

$$S_{eNodeB} = N_T \cdot N_{fig} \cdot B \cdot \gamma_{target} \tag{3.2}$$

In Eq. 3.2, N_T is the thermal noise power density calculated from Boltzmann's constant $k_B = 1.38 \times 10^{-23}$ J/K and the absolute temperature in Kelvin $T = 290$ K, to be – 174 dB/Hz. Moreover, N_{fig} is the eNodeB noise figure defining a degradation of γ due to RF circuitry components, calculated to be 2 dB for uplink [17, 18]. Finally, B is the RB bandwidth of 180 kHz. Substituting variables into Eq. (3.2), we get $S_{eNodeB} = -104.5 + \gamma_{target}$ dB. Proceeding with Fig. 3.11 and also considering an operator-predefined path loss at cell edge, γ_{target} could be properly estimated [19, 20]. M_{LNF} is the log-normal fading margin given by Jakes formula for a certain percentage of coverage and for specific environment (urban, dense urban, suburb, etc.). L_{BL} is the expected body loss considered either as 2 dB for handset palmtop or 0 dB for laptop:

$$L_{oper} = P_{T,s}^{UE,RB} - S_{eNodeB} - M_{LNF} - L_{BL} \Rightarrow$$

$$\gamma_{target}[dB] = 144.45 - L_{oper} - M_{LNF} - L_{BL} \tag{3.3}$$

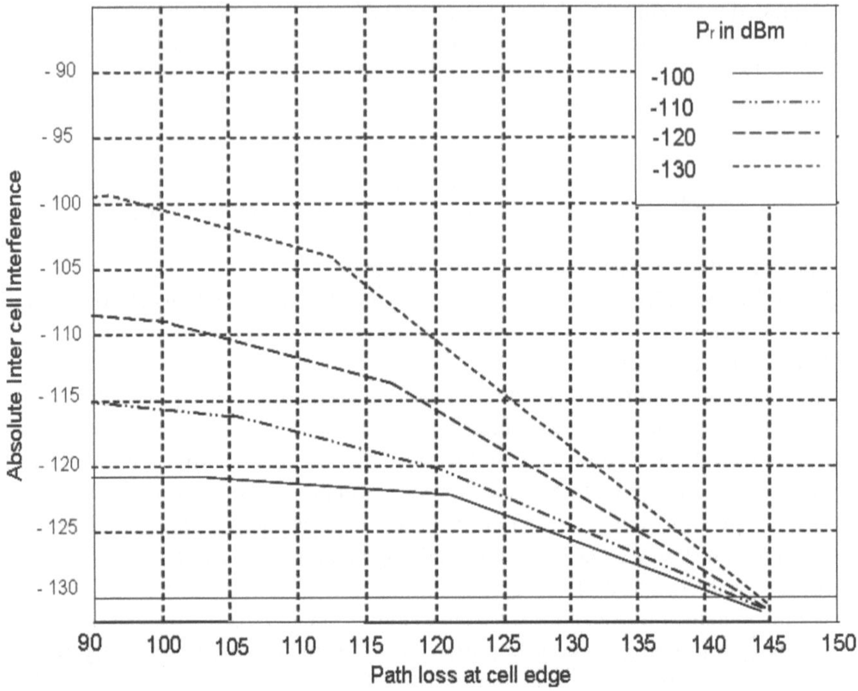

Fig. 3.11 Interference estimation versus path loss: drive tests results

- Estimate scheduler average number of uplink allocated RBs n_{RB}.
  Based on the target γ_{target} on cell edge, the number of allocated RBs n_{RB} is calculated considering uniform power distribution of nominal UE power P_{UE} over all transmitted RBs, as presented in Fig. 3.12.

$$\gamma_{target} = \frac{P_{UE}^{RB}}{noise + interference} = \frac{P_{UE}^{RB}/(L_{oper} \cdot n_{RB})}{(N_{RB} + I_{RB})} \Rightarrow$$
$$n_{RB} = \frac{P_{UE}^{RB}}{L_{oper} \cdot (N_{RB} + I_{RB}) \cdot \gamma_{target}} \tag{3.4}$$

- Estimate transmission rate per RB.
  To estimate the expected transmission rate per RB versus existing signal-to-noise ratio γ, real-drive test measurements have been analyzed. Since radio channel conditions are related to user velocity due to Doppler fadings, drive tests and the appropriate analysis have been executed for three different environments; EPA5 for pedestrians with average velocity of 5 km/h, EVA70 for in-car driving users with average velocity of 70 km/h, and ETU300 for high-speed users on highways [17]. Expected curves are presented in Fig. 3.13.

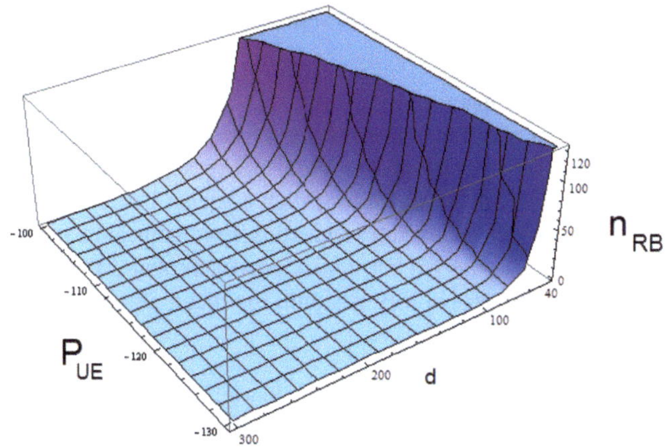

Fig. 3.12 Number of resource blocks (RBs) versus user equipment uplink power and distance to eNodeB

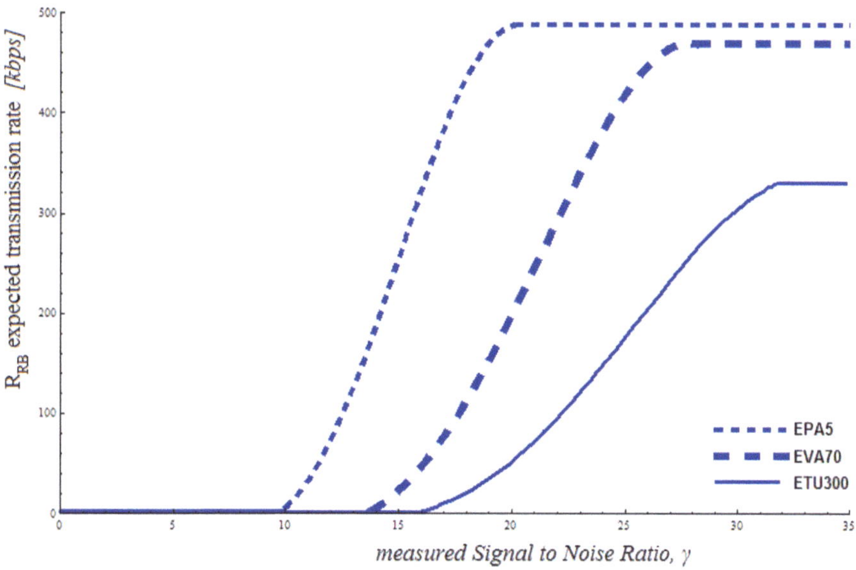

Fig. 3.13 Expected transmission rate per RB versus signal-to-noise ratio γ

- Calculate the expected total cell-edge transmission rate.
  Expected total transmission rate on cell edge is estimated by multiplying the average allocated number of resources n_{RB} from Eq. (3.4) with the expected transmission rate per RB from Fig. 3.13.

$$\langle R_{celledge} \rangle = n_{RB} \cdot R_{RB} \tag{3.5}$$

Average expected rate for VoIP service should be around 20 kbps, also considering retransmissions.

- Estimate expected retransmission rate
The average number of retransmissions m is a function of the expected packet error rate over physical OFDM layer. One MAC packet is considered to be corrupted if the BER is above a certain threshold; such a packet is retransmitted maximum v times before it is discarded and requesting retransmission from upper layers (RLC). Setting p to be the packet nonsuccessful probability (error probability) and assuming that this probability is small, the mean number of retransmissions can be calculated as:

$$m = p^{v} + \sum_{k=0}^{v-1} (k+1)p^{k}(1-p) = \frac{1-p^{v}}{1-p} \approx \frac{1}{1-p}, \quad p \ll 1 \tag{3.6}$$

Nonsuccessful probability is a function of MAC packet length M_{mac} and bit error probability p_b according to [16] is:

$$p = 1 - (1 - p_b)^{M_{mac}} \tag{3.7}$$

The average number of retransmissions is approximated as:

$$m \approx \frac{1}{1-p} = (1 - p_b)^{-M_{mac}} \approx (1 + M_{mac} \cdot p_b), \quad p_b \ll 1 \tag{3.8}$$

To estimate the expected bit error probability p_b, drive tests have been performed and the expected BER is calculated versus bit energy per noise spectral density, as presented in Fig. 3.14. The average size of MAC packet M_{mac} is difficult to estimate since its length is variable and explicitly decided upon MAC link adaptation, transport format selection functionality, and service, respectively. To proceed with drive tests, it is necessary to use network statistics from eNB. As an example, Ericsson eNodeB provides several statistical counters that could estimate the average packet length. Ericsson counter *pmUeThpVolUl [kb]* measures uplink MAC SDU volume and finally Ericsson counter *pmUeThpTimeUl [ms]* provides the period of MAC volume measurements in milliseconds, hence *pmUeThpVolUl/pmUeThpTimeUl*bits/1 ms provides average M_{mac}-transmitted bits per TTI.

- Estimate expected cell-edge throughput.
Expected throughput is easily calculated from Eqs. (3.5) and (3.8) as:

$$\langle T_{celledge} \rangle = \langle R_{celledge} \rangle \cdot \left(1 - \frac{m}{100}\right) = n_{RB} \cdot R_{RB} \cdot \left(1 - \frac{m}{100}\right) \tag{3.9}$$

- Estimate the average transmission delay.
To proceed with the expected delay, it is important to split the packet transmission delay into two parts. Consider the load $\rho = \lambda/\mu$ of the ratio of packet arrivals over

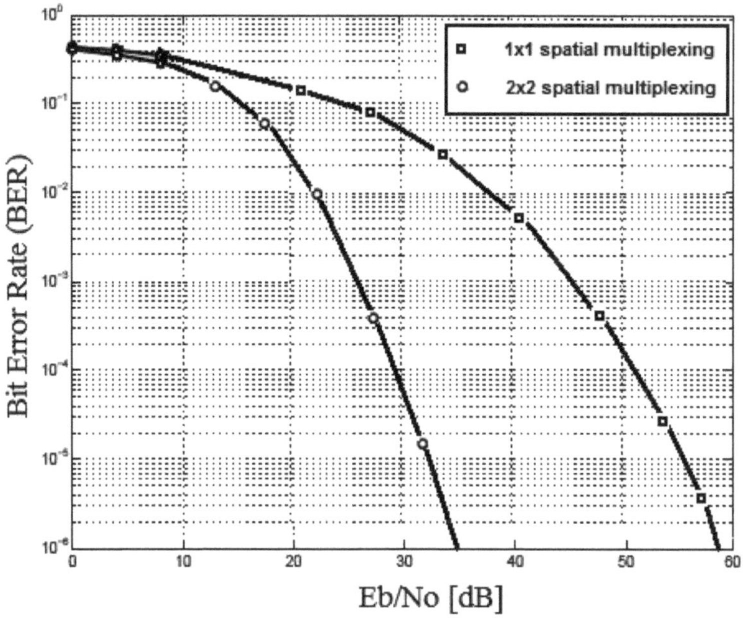

Fig. 3.14 Expected bit error rate (BER) versus bit received energy per noise spectral density

packet service time and in the general case suppose that specifically n packets exist in the system. Also, the signal processing delay per packet should be less or equal to one TTI = 1 ms. During first phase, a model for service layer IP source packet arrivals on UE, as they flow into the RLC/MAC layer waiting to be scheduled, is estimated [17] and presented in Fig. 3.15.

$$\begin{aligned} \bar{W} &= \sum_{n=1}^{\infty} n\pi_n \cdot TTI = \\ & \sum_{n=1}^{\infty} n \left[(1-\rho) \sum_{k=1}^{n} \left\{ (-1)^{n-k} e^{k\rho} \left[\frac{(k\rho)^{n-k}}{(n-k)!} \right] \right\} \right] \cdot TTI + \\ & \sum_{n=1}^{\infty} n \left[(1-\rho) \sum_{\substack{k=1 \\ k \neq n}}^{n} \left\{ (-1)^{n-k} e^{k\rho} \left[\frac{(k\rho)^{n-k-1}}{(n-k-1)!} \right] \right\} \right] \cdot TTI \end{aligned} \tag{3.10}$$

Second phase considers the MAC layer transmission, after uplink scheduling, where IP packets are forwarded into RLC/MAC layer for transmission over the physical medium. As expected from 3GPP standards and LTE protocols, IP packets will be segmented into many RLC/MAC-signaling data units (SDUs) to be mapped into OFDM RBs and transmitted over air interface. Multiple consecutive RBs n_{RB}

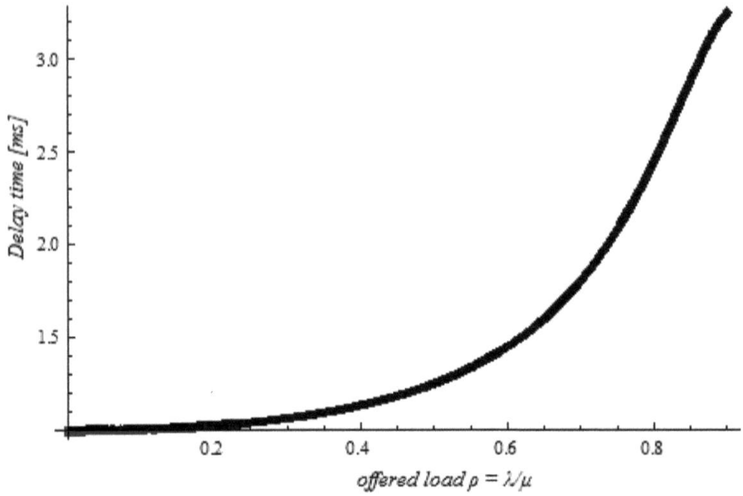

Fig. 3.15 Expected delay for service layer IP packet arrival on UE microprocessor

might be granted from scheduler for uplink transmission, minimizing the transmission latency and improving the UE throughput. Expected analysis will be based on transmissions of IP packets over RLC/MAC blocks based on channel conditions [16]. For a specific IP packet of average length M_I, the process defines fragmentation into $M_I + \lceil M_I/M_{mac} \rceil$ total number of RLC/MAC packets with $M_I + \lceil M_I/M_{mac} \rceil \cdot M_{over}$ total number of transmitted bits with fixed number of M_{over} header bits of 20 bytes per packet [20]. Considering also nonideal radio channel conditions, in such a scenario, the transmission time needed to completely transmit the IP packet will increase due to eventual retransmissions and nonscheduling periods of time. In general case, consider n_t number of uplink transmitted bits per RB (depending on MAC scheduler Link Adaptation Modulation and Coding Scheme), n_{RB} average allocated number of 180 kHz RB blocks per T_s transmission interval (as estimated in previous step), n_{AP} spatial multiplexing rank (MIMO or Tx diversity) and finally n and m integers indicating the average number of T_s units of time one MAC packet is not scheduled by scheduler and the average number of retransmissions one packet should undergo due to channel conditions, respectively (as calculated in previous step). Expected average whole IP packet transmission time would be:

$$W_{mac} = \frac{M_I + \lceil M_I/M_{mac} \rceil \cdot M_{over}}{n_{AP} \cdot n_{RB} \cdot n_t} \cdot TTI + (m+n) \cdot TTI \tag{3.11}$$

Uplink parameters n_t and n_{AP} could be easily calculated for cell-edge UEs considering that MAC scheduler will allocate QPSK modulation (2 bits per symbol) with Tx diversity, thus $n_{AP} = 1$. One subframe contains $14 \times 12 = 168$ REs where two OFMD symbols (24 REs) are devoted for sounding reference signals. Following Fig. 3.16, available number of uplink transmitted bits will be $n_{T_s} = (168 - 24) \times 2 =$

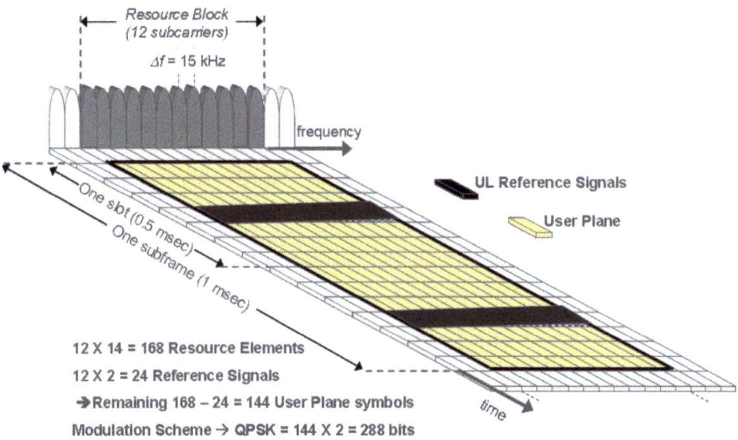

Fig. 3.16 Orthogonal frequency-division multiplexing (OFDM) user plane symbols for uplink transmission with quadrature phase shift keying (QPSK) action scheme

288 bits/ms. Regarding parameter n, using eNodeB-specific traffic counters from real traffic measurements, number of nonscheduled TTI intervals could be easily estimated. As an example, authors could propose Ericsson counters; following ratio *pmSessionTimeUe*/t_d, where *pmSessionTimeUe* is an Ericsson-based counter to measure the average total session service time of a UE considering transmissions and nonscheduled periods and t_d is the downloading measured time of a known size file from a server in a drive test. Average size of MAC packet M_{mac} has been already estimated in the previous steps. A good estimation of M_I is approximated using the ratio *PmPdcpVolUlDrb*/t_m, where *PmPdcpVolUlDrb* measures total uplink volume (PDCP SDUs) in an established Data Radio Bearer per measurement period in (kb) and t_m provides the measurement period. Considering both Eqs. (3.10) and (3.11), overall delay in the uplink transmission could be evaluated and compared with Fig. 3.10 3GPP QoS restrictions.

If during predefined step estimations, one step does not fulfill VoIP 3GPP standard restrictions or operator VoIP uplink determined requirements (throughput or BER), radio planner should reconsider (loose up some restrictions) cell size or path loss requirements and start recalculating cell size for smaller coverage conditions until all requirements and restrictions are fulfilled.

References

1. GSMA VoLTE. In GSMA VoLTE initiative. http://www.gsma.com/technicalprojects/volte. Accessed 2013.
2. Sauter, M. (2000). Voice over LTE via generic access (VOLGA)—A white paper. http://www.wirelessmoves.com. Accessed 2009.

3. 3GPP The Mobile Boradband Standard. http://www.3gpp.org/Technologies/Keywords-Acronyms/article/ims. Accessed 2010.
4. Ericsson (2013). What is voice over LTE. http://www.ericsson.com/res/thecompany/docs/corpinfo/volte_backgrounder.pdf. Jan 2013.
5. Tanaca, I., Koshimizu, T., & Nishida, K. (2009). CS fallback function for combined LTE and 3G circuit switched services. *NTT DOCOMO Technical Journal, 11*(3), 13–19.
6. 3GPP (2009). TR 25.913 feasibility study of evolved UTRA and UTRAN Rel-9.
7. Parkvall, D., & Beming, S. (2007). *3G evolution: HSPA and LTE for mobile broadband.* Oxford: Academic Press.
8. 3GPP TR 25.814 V7.1.0 (2006–2009). Physical layer aspects for evolved UTRAN terrestrial radio access (UTRA) release 7.
9. Technical Specification Group Radio Access Network (2009). Evolved universal terrestrial radio access (E-UTRA); LTE radio frequency (RF) system scenarios.
10. 3GPP (2008–2009). TS 36.942.
11. 3GPP (2006). TS 25.814 physical layer aspects for evolved universal terrestrial radio access (UTRA).
12. Elayoubi, S.-E., Ben Haddada, O., & Fourestie, B. (2008). Performance evaluation of frequency planning schemes in OFDM-based networks. *IEEE Transactions on Wireless Communications, 7*(5), 1623–1633.
13. Mao, X., Maaref, A., & Teo, K. H. (2008). *Adaptive soft frequency reuse for inter-cell isnterference coordination in SC-FDMA based 3GPP uplinks.* IEEE GLOBECOM Global Telecommunications Conference, pp. 1–6, Nov 30–Dec 4.
14. 3GPP TS 23.203 (2011). Policing and charging control architecture Rel-11, V11.4.0.
15. Pokhariyal, A., Kolding, T. E., & Mongensen, P. E. (2006). *Performance of downlink frequency domain packet scheduling for the UTRAN long term evolution.* IEEE 17th International Symposium on Personal Indoor and Mobile Radio Communications, pp. 1–5.
16. Louvros, S., Iossifides, A. C., Aggelis, K., Baltagiannis, A., Economou, G. (2012). *A semi-analytical macroscopic MAC layer model for LTE uplink.* Proceedings of 5th IFIP International Conference on New Technologies, Mobility and Security.
17. Louvros, S., Angelis, K., & Baltagiannis, A. (2011). *LTE cell coverage planning algorithm optimizing user cell throughput.* Proceedings of 11th IEEE International Conference on Telecommunications (ConTEL 2011), pp. 51–58.
18. Beniero, T., Redana, S., Hamalainen, J., & Raaf, B. (2009). *Effect on relaying on coverage on 3GPP LTE-advanced.* IEEE Vehicular Technology Conference, (VTC), pp. 1–5, April 26–29 2009.
19. Syed, A. B. (2009). *Dimensioning of LTE Network.* Dissertation Master Thesis. Department of Electrical and Computer Engineering, Helsinki University of Technology.
20. 3GPP TS 36.321 (2008). Evolved universal terrestrial radio access (E-UTRA); medium access control (MAC) protocol specification release 8, V8.1.0.
21. Bikos, A., & Sklavos, N. (2013, March/April). LTE/SAE security issues on 4G wireless networks. *IEEE Security and Privacy, 11*(2), 55–62.

Chapter 4
60 GHz Millimeter-Wave WLANs and WPANs: Introduction, System Design, and PHY Layer Challenges

Fotis Plessas and Nikolaos Terzopoulos

Abstract Millimeter-Wave technology enables multi-gigabit wireless communications, while a number of standards have been published targeting the license-free 60 GHz band. With much more spectrum available than the 2.4 GHz and 5 GHz bands, the 60 GHz band has wider channels, enabling higher data rates. On the other hand, this technology comes with many technical challenges due to the high carrier frequencies and the wide channel bandwidths used. In this context, an overview of the IEEE 802.11ad, the 802.15.3c, and the Wireless Gigabit Alliance (WGA) which are poised to define the next generation multi-Gbps wireless LANs and PANs will be given and the major challenges faced by the developers of this technology will be discussed. In order to emphasize on the PHY (RF front-end), critical building blocks such as: low-noise amplifier, mixer, frequency doubler/quadrupler, and VCO will be designed demonstrating that their performance against the relevant standards is satisfactory.

4.1 Introduction

License-free 60 GHz band makes it possible to achieve multi-gigabit connection throughput, and offers the ability to enable a variety of new uses and applications such as [1]:

1. The next generation TV link offering excellent quality video and audio for next generation of HD color depth and rates.
2. Download gigabytes of video or audio within seconds—send data between devices in the fraction of the time we need today.

F. Plessas (✉)
Department of Telecommunication Systems and Networks, Technological Education Institute of Messolonghi, Messolonghi, Greece
e-mail: fotis.plessas@gmail.com

N. Terzopoulos
Department of Computing & Communication Technologies, Faculty of Technology, Design & Environment, Oxford Brookes University, Wheatley, Oxford, UK

N. Sklavos et al. (eds.), *System-Level Design Methodologies for Telecommunication*, DOI 10.1007/978-3-319-00663-5_4,

3. Wireless computing—gigabit speed wireless I/O and untouched wireless display allows the user to remove the hard connection between computing platforms and peripherals without any compromising performance.

In all cases, the target is to reduce the use of cables.

4.1.1 Technical Characteristics and Specifications

4.1.1.1 Modulation and Data Rates

This section will refer to modulation schemes and date rates achieved [1]. The system should be scalable and should support both single carrier (SC) and Orthogonal frequency-division multiplexing (OFDM) modulation schemes and should also provide a low cost and low power solution to facilitate support to the wide variety of usages and applications envisioned for Multiple Gigabit Wireless Systems (MGWS) in both line-of-sight and non-line-of-sight channels. SC modulation includes the binary phase shift keying (BPSK), quadrature phase shift keying (QPSK) and the 16-quadrature amplitude modulation (QAM), offering data rates of around 4 Gbit/s whereas OFDM modulation includes the BPSK, QPSK, 16-QAM and 64-QAM, achieving data rates up to 7 Gbit/s.

4.1.1.2 Channel Access

The optimum channel access scheme used in 60 GHz operation should be the time division multiple access (TDMA), which is the most suitable when it comes to dealing with the challenges of the directional nature of communication and applications such as wireless displays. Finally, web browsing and file transfer should be supported via the contention-based access.

4.1.1.3 Beamforming

Beamforming is the key technology advantage in MGWS that is used in order to compensate for the additional 20 dB free space propagation loss in the 60 GHz band as compared to the 2.4/5 GHz bands [1, 2]. This propagation loss is due to the ultra high frequency nature of the 60 GHz band.

Beamforming makes use of multiple antenna elements to generate a beam of increased signal strength towards a certain direction. This beamforming gain is achieved by the use of the transmission of phase-shifted signals from multiple antenna elements. The signals are phase shifted so that they are added up comprehensibly at the target location (receiver end). As expected, the peak of the beamforming gain (Gb) increases as the number of antenna (Na) increases (i.e., Gb [dB] = 10log10Na). For example, a 16 element-antenna array can provide approximately 12 dB of peak

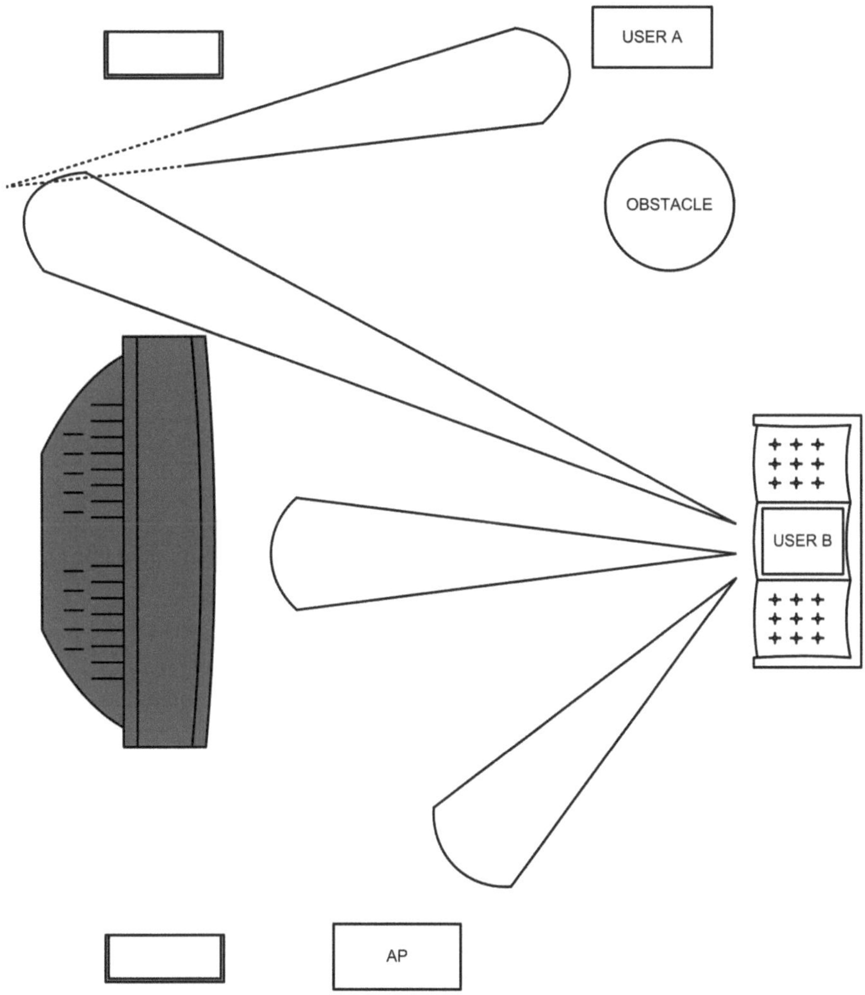

Fig. 4.1 An in-house example of beamforming

beamforming gain. If used at both the transmitter and the receiver, the 20 dB loss can be easily compensated for.

Compared to 2.4 and 5 GHz bands, beamforming is suitable for the MGWS since the 60 GHz band can allow us to pack many antenna elements in a very small area. For example, a square antenna array with 16 antenna elements (4×4 configuration) can be packed in only 1 cm^2 when adjacent antenna elements are separated by half wavelength. This is a crucial aspect considering the small form factor of the MGWS devices.

Figure 4.1 illustrates an example of how beamforming can be used within in an in-house environment. User A cannot communicate directly with user B and thus a non-line-of-sight link is used via reflection to establish connection between the two users. Therefore, the devices shown can communicate either through line-of-sight (direct contact) or non-line-of-sight links via reflections.

4.1.2 60 GHz Telecommunication Standards

The most common telecommunication standards used in the 60 GHz band are the IEEE P802.11ad, the IEEE802.15.3c, and the WGA. The following paragraphs describe the important specifications and the frequency allocation (channelization) for each scheme [1].

4.1.2.1 Overview of the IEEE P802.11ad Specifications

The IEEE P802.11ad specification provides a number of features that can meet the demands of the new usages and applications envisioned for MGWS, including [1, 3]:

- Data transmission rates up to 7 Gbit/s.
- Supplements and extends the 802.11 medium access control (MAC), supporting both scheduled access and contention-based access.
- Interoperability and communication at gigabit rates are guaranteed by enabling both low power and high performance devices.
- Supports beamforming, enabling noninterupting communication at distances beyond 10 m.
- Advanced security is supported by using advanced power management and the Galois/Counter Mode (GCM) of the advanced encryption standard (AES) encryption algorithm.
- Supports fast session transfer among 2.4, 5, and 60 GHz, which is known as multiband operation.
- The channel bandwidth is 2,160 MHz. Four center frequencies are defined at 58.32, 60.48, 62.64, and 64.8 GHz.

4.1.2.2 Overview of the IEEE 802.15.3c-20092 Specifications

In order to meet the demands of the usages and applications envisioned for MGWS, IEEE Std 802.15.3c-2009 provides features such as [1, 4]:

- Data transmission rates up to 5.8 Gbit/s.
- Supplements and extends the 802.15.3 medium access control (MAC), supporting both scheduled access, contention-based access, and frame aggregation for high throughput and quality of service.
- Interoperability and communication at gigabit rates is guaranteed by enabling both low power and high performance devices.

- Supports beamforming, enabling robust communication at distances of at least 10 m.
- In IEEE Std 802.15.3c, the channel bandwidth is 2,160 MHz. Four center frequencies are defined at 58.32, 60.48, 62.64, and 64.8 GHz.

4.1.2.3 Overview of the WGA Specifications

WGA visualizes a global wireless ecosystem of interoperable, high-performance devices that work together seamlessly to connect people in the digital age [1, 2]. This technology will enable multi-gigabit-speed wireless communications among these devices, and will drive industry to a single radio using the readily available, unlicensed 60 GHz spectrum. WGA specification includes key features to maximize performance, minimize implementation complexity, enable compatibility with existing WiFi and provide advanced security. Key features include:

- Supports for data transmission rates up to 7 Gbit/s meaning that all the devices based on the WGA specification should be capable to operate at gigabit rates.
- Supports low-power handheld devices such as cell phones, as well as high-performance devices such as PCs; includes advanced power management.
- Synchronized with IEEE 802.11; allowing native WiFi support and enabling devices to transparently switch between 802.11 networks operating in any of the following frequency bands: 2.4, 5, and 60 GHz.
- Supports enabling robust communication at distances beyond 10 m, beamforming and maximizing signal strength.
- Advanced security using the GCM of the AES encryption algorithm.
- Supports for high-performance wireless implementations of HDMI, DisplayPort, USB, SDIO, and PCIe through protocol adaptation layers (PALs).
- The channel bandwidth is 2,160 MHz. Four center frequencies are defined at 58.32, 60.48, 62.64, and 64.8 GHz.

WGA's Physical and MAC layer specifications are harmonized with those of IEEE 802.11ad version Draft 2.0.

4.2 Physical Layer—RF Front-End Design

In order to fulfill the aforementioned specifications, we present the system level design of a transparent RF front-end which is the key component of the PHY layer. The RF front-end targets the integration of the critical mm-wave circuits (Low-noise amplifier (LNA), downconverter, upconverter, and power amplifiers (PAs)) along with a super-heterodyne architecture with an intermediate frequency (IF; e.g., at 5 GHz band), modulator/demodulator and phase-locked loops (PLL), a lowpass baseband filter, and a bandpass IF filter [5]. An mm-wave PLL is designed by quadrupling a 15 GHz voltage-controlled oscillator (VCO) in order to drive the up/down converters with a specific phase noise profile [5]. Mixed signal functions (ADC/DAC)

Table 4.1 Indicative requirements for the receiver, the transmitter, and the PLL

	Receiver	Transmitter	PLL
NF (dB)	7.0		
P_{1dB} (dBm)	− 18		
Gain (dB)	11		
Phase noise (dBc/Hz)			− 90
Output power		+ 10 dBm	

NF noise figure

could also be integrated, being the input/output of the super-heterodyne architecture, interfacing with the rest of the baseband modem.

I/Q modulation/demodulation can be realized into either the analog (RF front-end) or the digital domain (baseband). When implemented into the analog domain, a number of inherent errors such as the gain balance, quadrature-phase balance, DC offsets, and impedance matching can degrade the performance. Recent advances in high-speed analog-to-digital converters allow the I/Q modulator/demodulator to be implemented digitally. In order to keep the complexity within the RF front-end we are adopting the I/Q modulator/demodulator implementation employed within the RF front-end.

All analog, RF, mixed signal, and mm-wave transceiver functions are designed using a well established 90 nm CMOS process with a 1.2 V core power supply, and the key blocks will be discussed in the following sections.

Finally, to meet the specifications, regardless of the selected architecture, modeling and design should be done in parallel for all the components consisting the major system blocks. Apart from using electronic design automation (EDA) computer-aided design (CAD) tools such as Cadence Virtuoso©, Agilent ADS©, and 3D EM software for the mm-wave blocks, custom models should be generated for the critical active as well as for the passive components.

Some indicative important requirements for the receiver, the transmitter, and the PLL are presented in Table 4.1 (please note that these indicative figures are likely to differ depending on the targeted standard and application).

4.3 RF Front-End Blocks/Circuits

Using the aforementioned requirements we can extract the specifications for each one of the analog front-end blocks: receiver, transmitter, and the PLL. In the following sections, the design of the LNA, the downconverter, the upconverter, the preamplifier, the VCO, and the quadrupler will be discussed.

Provided that there is a passband filter with 1 dB insertion loss just before the LNA (Fig. 4.2), the gain, the NF, and the P_{1dB} for the LNA and the mixer can be calculated. Thus, the design specifications of the LNA are: NF of 5.3 dB, P_{1dB} of − 9 dB, and a gain of 12 dB. Likewise, the mixer specifications are: NF of 10 dB, P_{1dB} of − 5 dB, and a gain of 0 dB.

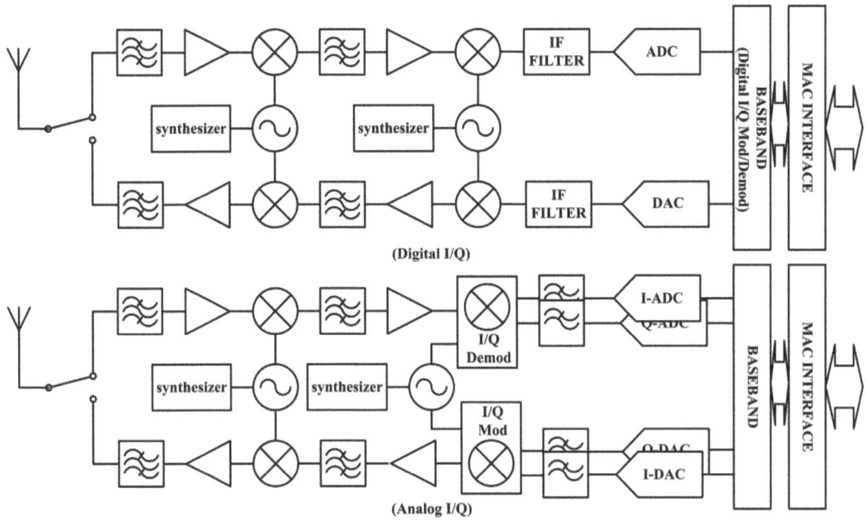

Fig. 4.2 RF Front-end using analog or digital I/Q

4.3.1 Receiver

4.3.1.1 Low-Noise Amplifier (LNA)

LNA is probably the most crucial subcircuit of the receiver because it greatly defines the total noise figure of the entire system.

The specifications given, currently dictate that at least two stages of amplification are needed. As it is expected, the first amplification stage needs to be optimized to achieve low noise contribution whereas the second stage deals with the design aspects like gain and linearity. Therefore, the LNA presented further was developed based on the aforementioned principle.

A good candidate for the first amplification stage would be the common source configuration with an inductive source degeneration and inductive load as shown in Fig. 4.3 [6, 7].

A common practice while designing a low noise amplifier and in order to avoid undesirable oscillations while achieving optimum isolation between the LNA's input and the mixer's output is that in all cases the LNA input should be isolated from its output node. This could be achieved by using the cascode LNA configuration. In this configuration, we have two transistors vertically stacked, having the issue here of not being suitable for low voltage applications due to the limited head room. To overcome the low voltage issue the folded cascode configuration is proposed. The DC isolation between the two stages of the cascode amplifier makes it suitable for low voltage applications. The drawback of this topology is the use of a second stage of a common gate configuration being used as the output stage with additional power consumption.

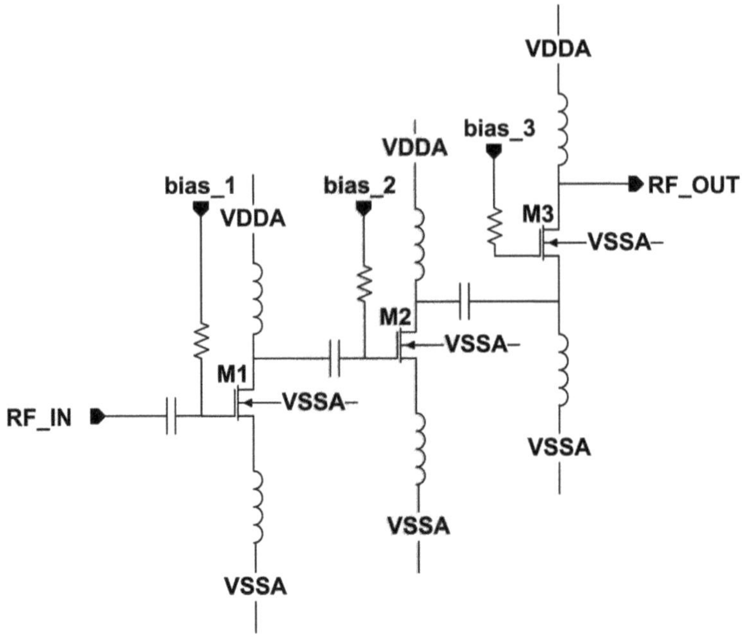

Fig. 4.3 The proposed LNA

As shown in the schematic diagram, L (coplanar transmission lines) and C matching networks are connected between all three stages for the necessary impedance matching. Worst case post-layout simulation results have indicated that the proposed design can easily achieve an NF of 5 dB, a P_{1dB} of − 11 dB, and an S11 of − 18 dB, providing a gain of 13 dB while consuming 13 mA from a 1.2 V supply at 60 GHz.

4.3.1.2 Downconverter

The main purpose of an RF mixer is to downconvert the input high frequency signal (for e.g., 60 GHz) to a lower frequency (for e.g., 5 GHz). A typical mixer topology consists of a differential mixing pair, a local oscillator (LO) amplifier, and an output buffer. A single balanced mixer topology can be implemented using a passive differential mixing pair (Fig. 4.4); this type of mixer exhibits good linearity and simplicity in its design. However, the drawback here is the insufficient signal isolation between the mixer's input and output. In our case, though, this does not appear to be of a great importance since this signal is filtered out by the resonating output load [6].

In order to ensure minimum transition times from one state to another, the gates of the switching transistors are biased at the threshold voltage, achieving low noise performance with minimum losses for a given LO's input signal (power level).

The performance of a passive mixer greatly depends on the signal from the LO. Hence, this signal must be further amplified in order to improve the mixer's gain,

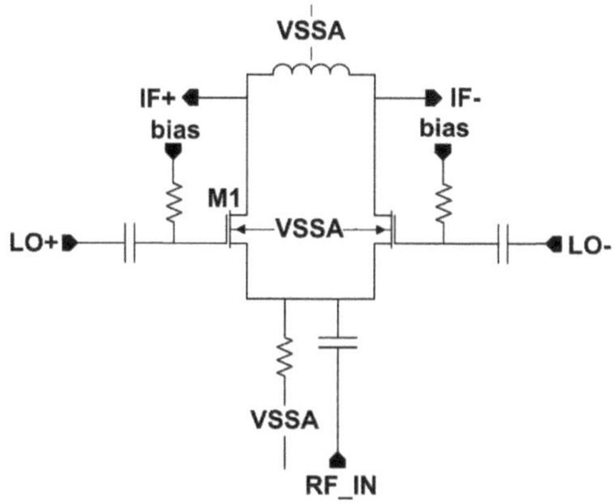

Fig. 4.4 The mixing pair

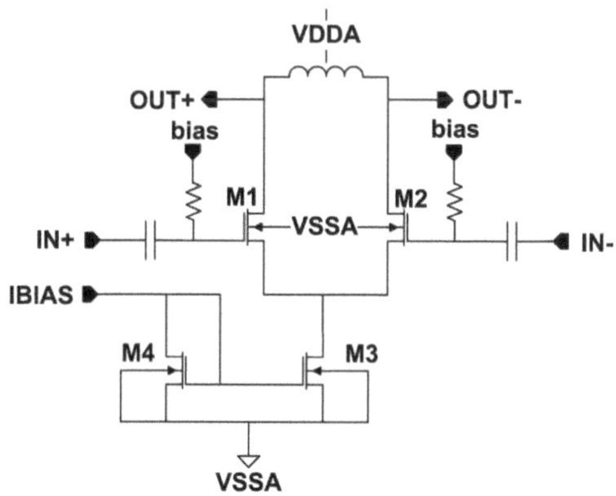

Fig. 4.5 The schematic of the LO amplifier

noise, and linearity. A signal amplifier, based on the differential pair configuration with a tail current source can be used in order to produce an adequate voltage level at the gates of the switching transistors (Fig. 4.5).

To ensure that the mixer is well isolated from the load, a buffer (source follower configuration) can be employed, connected at the mixer's output. The reasons behind the selection of this topology is twofold: firstly, the mixer's output will probably need to be connected to a 50 Ω resistor which will not be sufficiently driven by the differential mixing pair and secondly, the source follower will isolate the mixer from the environmental interferences.

Worst case post-layout simulation results meet the original design requirements. The NF is 9.5 dB, the P_{1dB} is −7 dB and the gain is 0 dB. The circuit's power consumption is approximately 20 mA.

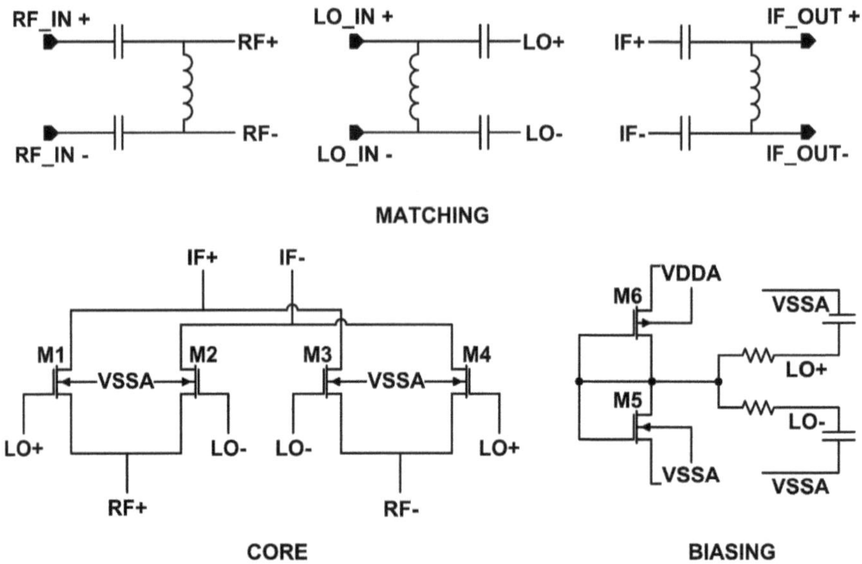

Fig. 4.6 The complete schematic of the mixer

4.3.2 Transmitter

4.3.2.1 Upconverter

The input signal from the IF (e.g., 5 GHz) should be upconverted to the 60 GHz band. The upconverter consists of a mixer and a preamplifier (buffer) to compensate for the mixer's losses [6]. An external or internal power amplifier should be used to reach the maximum output power of + 10 dBm.

In Fig. 4.6, the schematic of the proposed ring mixer is shown. The mixing procedure involves the switching polarity technique. The matching network is realized using inductors and capacitors. In order to provide sufficient isolation, this matching network acts as a virtual ground for the LO input and RF output when connected to IF input, and for the IF input when connected at the RF output.

The odd order intermodulation products are being filtered out while an advantage of this specific mixer topology is the rejection of the intermodulation products in general, provided it is well balanced. To achieve this, the physical design of the mixer should be done in such a way as to reduce the parasitic elements to the minimum. Other important mixer characteristics include the 1 dB compression point and the low noise figure compared with other Gilbert-type double-balanced mixers.

Post-layout simulation results gave a conversion loss of − 6.5 dB, and a 0 dBm P_{1dB}.

The recommended preamplifier's architecture could be implemented using a three stage amplifier. Each stage has a cascode topology for good isolation between the

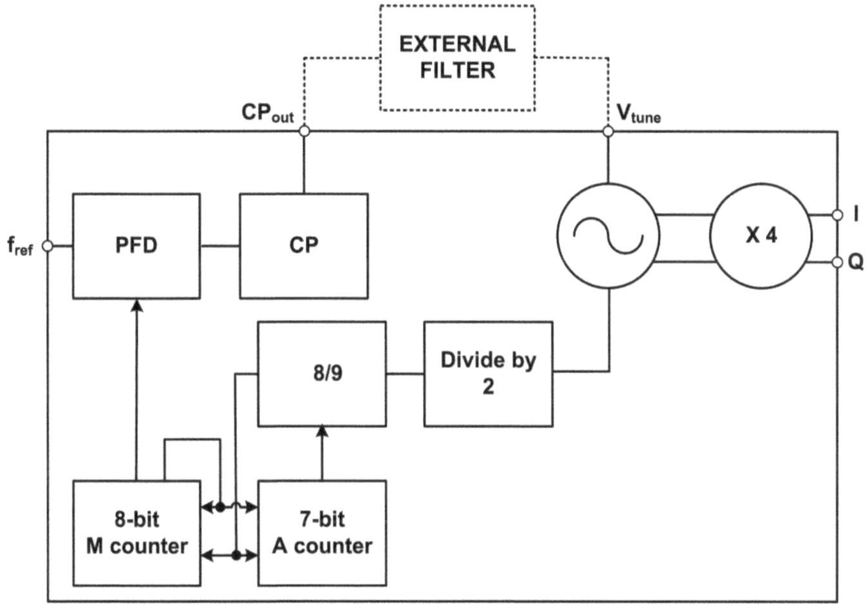

Fig. 4.7 The proposed PLL architecture

input and the output and is designed for optimum linearity. Strip lines can be used to tune the circuit while its gain can be varied by altering the bias voltage of the cascade transistors [6, 7]. The preamplifier achieves a gain which is 8 dB higher than that required to compensate for the losses of the passive mixer.

4.3.3 Phase-Locked Loop (PLL)

A quadrature differential PLL for the license-free 60 GHz band, with a frequency step of 50 MHz, suitable for wireless transceivers is discussed in this section. The proposed architecture, shown in Fig. 4.7, allows the use of a lower-frequency PLL (i.e., 15 GHz), an approach which is very advantageous towards implementing a millimeter wave (i.e., 60 GHz) wide-tuning and low-phase-noise PLL [8, 9].

The PLL consists of a 15 GHz quadrature differential VCO (QVCO), a programmable charge pump (CP), a high frequency divide-by-2 divider, a pulse-swallow divider including an 8/9 prescaler, a phase frequency detector (PFD), a quadrupler, a bandgap reference (BGR), and control logic. Finally, the quadrapler is a combination of a 15–30 GHz doubler, two 30 GHz amplifiers, a polyphase filter, a 30–60 GHz doubler, and two 60 GHz amplifiers. Within this context, only the VCO and the frequency quadrupler, the two most critical blocks are discussed further [10].

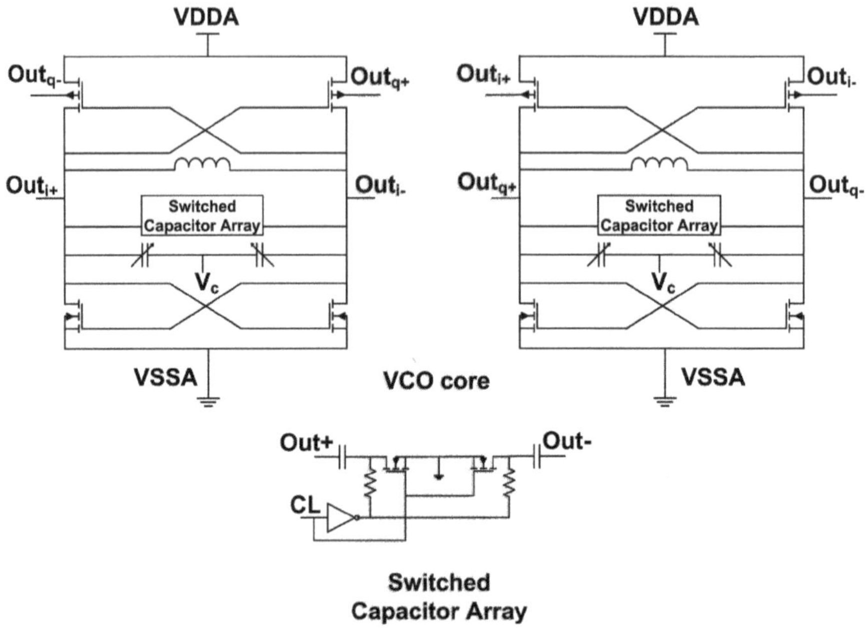

Fig. 4.8 The schematic of the VCO

4.3.3.1 Voltage-Controlled Oscillator (VCO)

An LC CMOS quadrature cross-coupled VCO appears to be the most suitable design choice for 15 GHz region operation [10]. The use of both nMOS and pMOS transistors leads to a better noise response. At such high frequencies, varactors with extremely high tuning range have very low Q factor at the low capacitance side. This prompts the use of a dual-tuning model as in [11] employing switched capacitor arrays. The theoretical Q value of these arrays is close to the Q value of the capacitor used, while the transistor switching affects the resulting performance considerably by imposing more restraints in the on/off capacitance ratio. The schematic of the implemented VCO is shown in Fig. 4.8.

It is a typical LC CMOS cross-coupled design with a switched capacitor array for coarse tuning. The varactors used are pMOS transistors on p substrates available in the process. For visual clarity the quadrature coupling network has been omitted and the AC coupling directly appears in the figure. For the quadrature signal generation, two distinct differential VCOs are used in conjunction with back-gate coupling through the pMOS body. The coupling network is a capacitive network with a resistive connection to the supply, for bulk biasing purposes. Back-gate coupling, though slower in achieving the 90° phase difference, has a lower impact on phase noise than coupling through transistors in parallel with the cross-coupled pair. For the coarse tuning, nMOS transistors are used as switches. In order to optimize the switches' behavior on both states, the internal node DC voltage is dynamically adjusted with

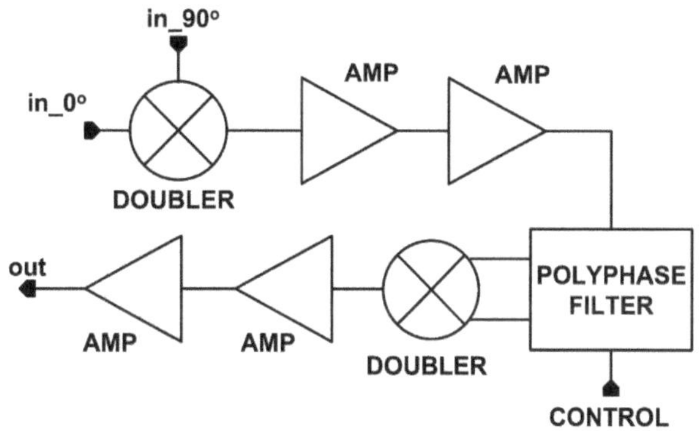

Fig. 4.9 The full quadrupler topology

respect to the control voltage, similarly as in [12]. This allows for a minimum off capacitance since the p–n junction of the drain is at high reverse voltage, resulting in the least capacitance. When the nMOS is switched on, the DC operation point goes to zero achieving zero DC current and minimum on resistance.

In order to drive the divider, a current-mode logic (CML) buffer is employed. The buffer is a two-stage design with the second stage offering dual outputs to feed both stages.

The simulation results demonstrate that the VCO oscillates from 13 GHz to 14.9 GHz. The achieved phase noise at an offset of 1 MHz from the carrier is − 112.5 dBc/Hz. The power consumption of the core is 12.7 mW.

4.3.3.2 Quadrupler

The conversion from 15 GHz to 60 GHz is achieved by the use of a frequency quadrupler topology as shown in Fig. 4.9. It consists of two frequency doublers, one polyphase filter, gain amplifiers, and the bias circuit [13].

Regarding the type of the frequency doublers, a passive double-balanced mixer can be chosen (Fig. 4.6). Its main advantage in comparison to other topologies is its higher linearity.

The disadvantages of this doubler is the requirement for a high-level signal into the LO input whereas at the same time it exhibits high conversion loss. As a result, the output signal must be amplified before it drives the following stages. Furthermore, it should be taken into account that this output will drive a second frequency doubler, which also requires high input level. Thus, two gain amplifiers have to be employed after the first frequency doubler. Specifically, two LO amplifiers have been used in cascade to amplify the doubler output resulting in a gain of about 15 dB (Fig. 4.5). The next step is to again double the frequency from 30 GHz to 60 GHz. However, such a doubler would require two differential signals of the same frequency but shifted by 90°. Therefore, a polyphase filter [14, 15] is inserted before the doubler, providing two 90°-shifted signals at 30 GHz frequency (Fig. 4.10).

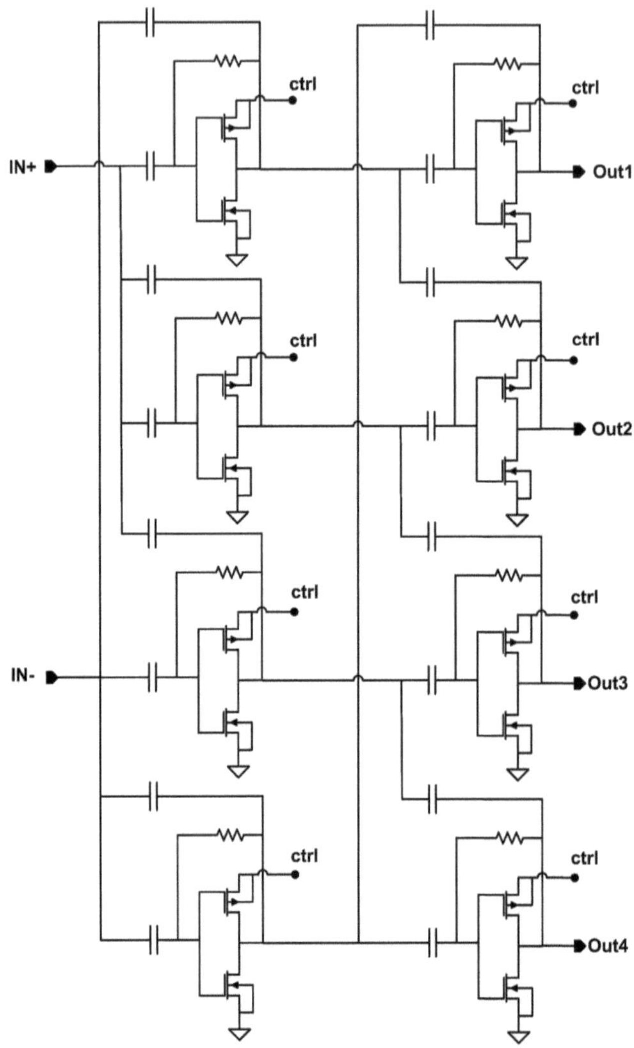

Fig. 4.10 Active polyphase filter

To avoid the additional use of another polyphase filter before the first doubler, the two shifted signals are provided by the quad output VCO, thus reducing the number of the polyphase filters needed. An active type of polyphase filter was used offering the capability for fine regulation in the case that the phase shift deviates from the required value. The second doubler follows the polyphase filter, resulting in a frequency doubling from 30 GHz to 60 GHz. Again, its output is considerably attenuated and must be amplified. Therefore, two more gain LO amplifies are used to drive the signal.

4.4 Conclusions

In this chapter, the main indoor wireless telecommunication standards for the licensed-free area of 60 GHz were presented. Emphasis was laid on the physical layer and mainly in the topology selection of the blocks consisting of an analog front-end suitable for wireless applications. The most critical blocks such as the LNA, the mixers, the VCO, and the PLL were thoroughly discussed and the recommended architectures were presented for each case in accordance to the specifications of the wireless system. The technology on which the simulation results are based, is a well-established CMOS 90 nm process for the proof of concept, but other mainstream processes can also be used in the same efficient way.

References

1. ITU (2011). Report ITU-R M.227, Multi Gigabit Wireless Systems in frequencies around 60 GHz. http://www.itu.int/dms_pub/itu-r/opb/rep/R-REP-M.2227-2011-PDF-E.pdf. Accessed online 20 May 2013.
2. Wireless Gigabit Alliance (2010). Defining the future of multi-gigabit wireless communications. http://wirelessgigabitalliance.org/. Accessed 20 May 2013.
3. IEEE (2012). IEEE Std 802.11ad-2012, IEEE Standard for Information technology—Telecommunications and information exchange between systems—Local and metropolitan area networks—Specific requirements-Part 11: Wireless LAN medium access control (MAC) and physical layer (PHY) specifications Amendment 3: Enhancements for very high throughput in the 60 GHz band. http://ieeexplore.ieee.org/servlet/opac?punumber = 6178209. Accessed 20 May 2013.
4. IEEE (2009). IEEE Std 802.15.3.c-2009, IEEE Standard for Information technology—Telecommunications and information exchange between systems—Local and metropolitan area networks—Specific requirements-Part 15.3: Wireless medium access control (MAC) and physical layer (PHY) specifications for high rate wireless personal area networks (WPANs), Amendment 2: Millimeter-wave-based alternative physical layer extension. http://ieeexplore.ieee.org/servlet/opac?punumber = 5284442. Accessed 20 May 2013.
5. Makri, R., et al (2011). *Next generation millimeter wave backhaul radio: Overall system design for GbE 60 GHz PtP wireless radio of high CMOS integration*. Paper presented at the 18th IEEE International Conference on Electronics, Circuits, and Systems, Beirut, Lebanon, 11–14 December 2011.
6. Simitsakis, P., Liolis, S., Psyllos, D., Mountrichas, L., Sotiriadis, P.-P. (2011). *Design of a 1.2-V 60 GHz transceiver in a 90 nm CMOS RF technology*. Paper presented at the 18th IEEE International Conference on Electronics, Circuits, and Systems, Beirut, Lebanon, 11–14 December 2011.
7. Yao, T., Gordon, M. Q., Tang, K. K. W., Yau, K. H. K., Yang, M. T., Schvan, P., Voinigescu, S. P. (2006). Algorithmic design of CMOS LNAs and PAs for 60-GHz Radio. *IEEE Journal of Solid State Circuits, 41*(1), 17–22.
8. Musa, A., et al (2010). *A 58–63.6 GHz Quadrature PLL Frequency Synthesizer in 65 nm CMOS*. Paper presented at the 2010 IEEE Asian Solid State Circuits Conference (A-SSCC), Beijing China, November 2010.
9. Gunnarsson, S. E., et al (2007). 60 GHz single-chip front-end MMICs and systems for multi-Gb/s wireless communication. *IEEE Journal of Solid State Circuits, 42,* 1143–1157.
10. Plessas, F., Panagiotopoulos, V., Kalenteridis, V., Souliotis, G., Liakou, F., Koutsomitsos, S., Siskos, S., Birbas, A. (2011). *A 60-GHz Quadrature PLL in 90 nm CMOS*. Paper presented in the

18th IEEE International Conference on Electronics, Circuits, and Systems, Beirut, Lebanon, December 11–14 2011.
11. Kim, H.-R., Oh, S.-M., Kim, S.-D., Youn, Y.-S., Lee, S.-G. (2003). *Low Power Quadrature VCO with the Back-Gate Coupling*. Paper presented at the 29th European Solid-State Circuits Conference (ESSCIRC), Estoril, Portugal, 699–701, September 2003.
12. Zhang, K., Cheng, S., Zhou, X., Li, W., Liu, R. (2009). A wideband differentially switch-tuned CMOS monolithic quadrature VCO with a low Kvco and high linearity. *Microelectronics Journal, 40*(6), 881–886.
13. Souliotis, G., Plessas, F., Liakou, F., Birbas, M. (2012). A 90 nm CMOS 15/60 GHz frequency quadrupler. *International Journal of Electronics*. doi:10.1080/00207217.2012.751321.
14. Behbahani, F., Kishigai, Y., Leete, J., Abidi, A. (2001). CMOS mixers and polyphase filters for large image rejection. *IEEE Journal of Solid-State Circuits, 36,* 873–887.
15. Tillman, F., Sjoland, H. (2001). A polyphase filter based on CMOS inverters. *Analog Integrated Circuits and Signal Processing, 50,* 7–12.

Chapter 5
Modeling the Operation of CMOS Primitive Circuits and MOSFET Devices

Labros Bisdounis

Abstract Estimation of complementary metal-oxide semiconductor (CMOS) circuits' behavior, in terms of analysis and computation of their dynamic characteristics (such as propagation delay, transition time, and energy dissipation) is today a standard part of digital circuit design. Since these characteristics are critical design parameters in CMOS digital circuits, much effort has to be devoted for the extraction of accurate, analytical expressions for primitive circuits. Using transistor-level simulators with continuous-time modeling of the devices such as SPICE can be very expensive in terms of storage and computation time. Hence, much of the research has addressed the development of analytical timing and energy dissipation models, without the necessity of expensive numerical iterations. This chapter mainly regards the methodology for the derivation of closed-form, accurate expressions for the aforementioned parameters. The operational conditions of primitive CMOS structures are determined and the differential equations describing their operation are solved analytically by using appropriate approximations in order to simplify the modeling procedure, without significant influence in the accuracy. As a case study, the CMOS inverter is used. Following a detailed analysis of the inverter operation, accurate expressions for its output response are derived for the different operation regions, and based on this analysis, analytical expressions for the calculation of the timing and energy parameters can be produced. The derived models account for the influences of input voltage transition time, device sizes, parasitic capacitances, output load, as well as small-geometry device effects. The inverter model can be extended to multi-input CMOS gates by using reduction techniques of series-connected and parallel-connected transistors. Since the accuracy of the used MOSFET device I–V model determines the accuracy of primitive circuits' timing and energy models to a large extent, accurate and compact device models that take into account the influences of predominant effects in modern nanometer device technologies should be adopted.

L. Bisdounis (✉)
Electrical Engineering Department, Technological Educational Institute of Patras,
1, M. Alexandrou Street, 263 34 Patras, Greece
e-mail: bisdounis@teipat.gr

N. Sklavos et al. (eds.), *System-Level Design Methodologies for Telecommunication*,
DOI 10.1007/978-3-319-00663-5_5,

5.1 Introduction

Switching speed is traditionally a critical performance parameter in circuits and systems [1]. In addition, energy dissipation is a very critical parameter that has to be taken into account during circuit design, to avoid problems related to heating in high-performance applications and those related to energy savings in battery-oriented portable applications [2].

Thus, computer-aided design tools should include efficient methods for fast and accurate computation of such parameters, in terms of accurate, analytical expressions for timing and energy models of basic circuits [3–7]. From the practical use point of view, improved analytical timing and energy dissipation models can be applied to various types of simulation (i.e., fast-timing simulation, switch-level and gate-level timing, and mixed-mode simulation), subcircuits' timing/energy analysis and characterization, as well as to transistor sizing. Examples of application of analytical models in fast-timing and gate-level simulation are the ILLIADS [8, 9] and HALOTIS [10] tools, respectively. The main advantage of the analytical modeling approach is that it does not require presimulation as in tabular methods or empirical equation techniques, in order to improve the simulation speed. On the other hand, since analytical models for circuit primitives (like the CMOS inverter) preserve the nonlinearity of the devices, they incorporate the impact of modern technologies' effects resulting in high accuracy, without the need for computational expensive iterations and numerical methods used in SPICE-like circuit simulators.

The methodology for analytical modeling of CMOS primitive circuits' operation is constituted from the following main steps:

- Understanding of the circuit operation.
- Creation of a circuit model including the parasitics to be considered.
- Derivation of the differential equation that describes the circuit operation, i.e., application of the Kirchhoff's current law at the output node of the primitive circuit.
- Definition of the operating regions of the primitive CMOS circuit's transistors in every time interval of the circuit operation.
- Adoption of an appropriate current model for the transistors in the main operating regions (cutoff, linear, saturation). The adopted transistor model should match the I–V characteristics of the transistors by accounting first-order and main second-order effects [11, 12], according to the used technology process. The transistor current model should combine simplicity to provide the ability for derivation of explicit expressions for design parameters, and accuracy to account for the influences of main device physical mechanisms.
- Definition of the circuit operational conditions, which determines the bounds of different operating regions.
- Derivation of the differential equation analytical solution in each region of operation, in order to determine the output voltage waveform of the circuit. The output voltage waveform is determined as an integration of the device currents in contrast to the average current method where the current is assumed equal to the average of its values at the limits of the time interval of interest.

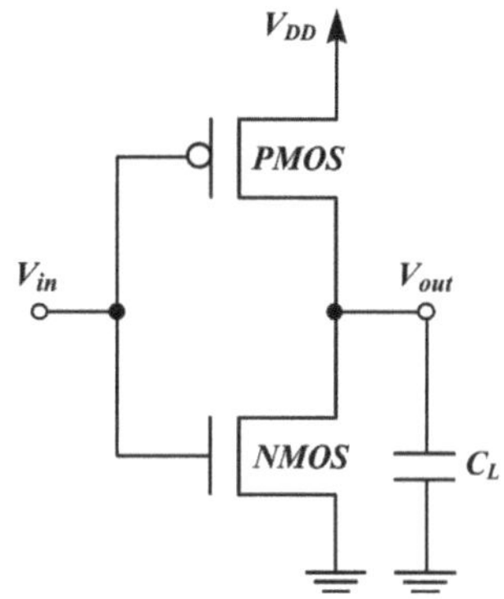

Fig. 5.1 The CMOS inverter

- Since the target is to combine simplicity and accuracy, the methodology should include creative and reasonable assumptions in case the differential equation cannot be solved analytically. Circuit simulations can be used to define candidate regions for approximations.
- Use of smart techniques (such as Taylor series expansions) to solve boundary equations in order to avoid numerical approaches.
- Adoption of efficient techniques for the reduction of complex circuits (i.e., multi-input CMOS gates) to primitive circuits (i.e., CMOS inverter) [6, 13, 14].

5.2 Analytical Modeling of CMOS Primitive Circuits

In order to apply the aforementioned methodology to a case study primitive circuit, the CMOS inverter (Fig. 5.1) and its energy dissipation are considered. The importance of modeling the timing response and the energy dissipation of CMOS inverters comes from the fact that the clock distribution networks and busses in digital integrated circuits are based on inverters or inverter-like circuits, which have to be carefully modeled since these circuits account for a great fraction of circuits' delay and energy dissipation [15]. Another equally important reason to obtain accurate models for the inverter delay and energy dissipation is that several methods for reducing multi-input CMOS gates to equivalent inverters have been proposed [6, 13, 14].

The energy dissipation of a CMOS inverter is constituted from the dynamic energy dissipation due to charge and discharge of the output load during the inverter switching, the leakage energy dissipation, and the short-circuit energy dissipation due to the direct current path from power supply to ground during the inverter switching [1]. During the rising transition of the output node, half of the supplied energy is stored in the load capacitance (and will be lost to the ground during the falling output transition), and the rest is lost in the form of heat at the PMOS device. The supplied energy and consequently the dynamic part of energy dissipation are independent from the output waveform of the inverter [1]. The leakage energy dissipation is caused by the reverse-bias diode leakage current at the device diffusion areas, the subthreshold current through the turned-off transistor channel, and the tunneling current through

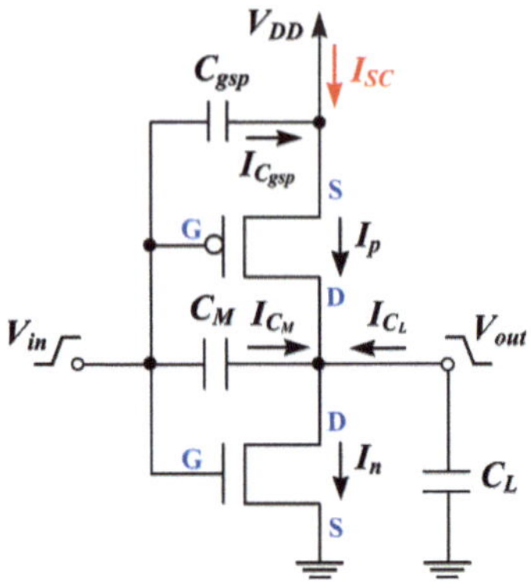

Fig. 5.2 CMOS inverter model used for the computation of the short-circuit energy dissipation

the very thin gate oxide. This type of energy dissipation is mainly dependent on the technology process parameters and it is significant for nanometer transistors [1, 2].

The short-circuit energy dissipation depends on the transition time of the inverter input voltage transition time, the output load, the supply voltage, and the internal design characteristics of the inverter [6, 16]. Hence, the computation of this type of energy dissipation is complex, and accurate analysis of the output voltage waveform is required by following the steps of the aforementioned methodology.

For the computation of the inverter short-circuit energy dissipation, analytical expressions of the output waveform in the required operating regions, are derived. These expressions take into account the current through both transistors and the influence of the gate-drain coupling capacitance, and they are valid for a wide range of input transition times, output loads, supply voltages, and device sizes. In addition, the influence of the short-circuiting transistor's gate-source capacitance is included when computing the short-circuit energy dissipation of the inverter. The circuit model that includes the parasitics to be considered is shown in Fig. 5.2.

The derivations presented further are for a rising input ramp: $V_{\text{in}} = V_{\text{DD}} \cdot (t/\tau)$ for $0 \leq t \leq \tau$, $\text{V}_{\text{in}} = 0$ for $t \leq 0$, and $V_{\text{in}} = V_{\text{DD}}$ for $t \geq \tau$, where τ is the input rise time. The analysis for a falling input is symmetrical. The differential equation describing the discharge of the load capacitance C_{L} for the CMOS inverter (Fig. 5.2), taking into account the current through the gate-drain coupling capacitance, (C_{M}) is written as [3]:

$$C_{\text{L}}\frac{\text{d}V_{out}}{\text{d}t} = C_{\text{M}}\left(\frac{\text{d}V_{\text{in}}}{\text{d}t} - \frac{\text{d}V_{\text{out}}}{\text{d}t}\right) + I_{\text{p}} - I_{\text{n}} \tag{5.1}$$

After normalizing voltages with respect to the supply voltage (V_{DD}), and using the variable $x = t/\tau$, the differential Eq. (5.1) is written as

$$\frac{\text{d}u_{\text{out}}}{\text{d}x} = c_{\text{m}} + \frac{(I_{\text{p}} - I_{\text{n}})\tau}{(C_{\text{L}} + C_{\text{M}})V_{\text{DD}}}, \tag{5.2}$$

where

$$c_{\text{m}} = \frac{C_{\text{M}}}{C_{\text{L}} + C_{\text{M}}} \tag{5.3}$$

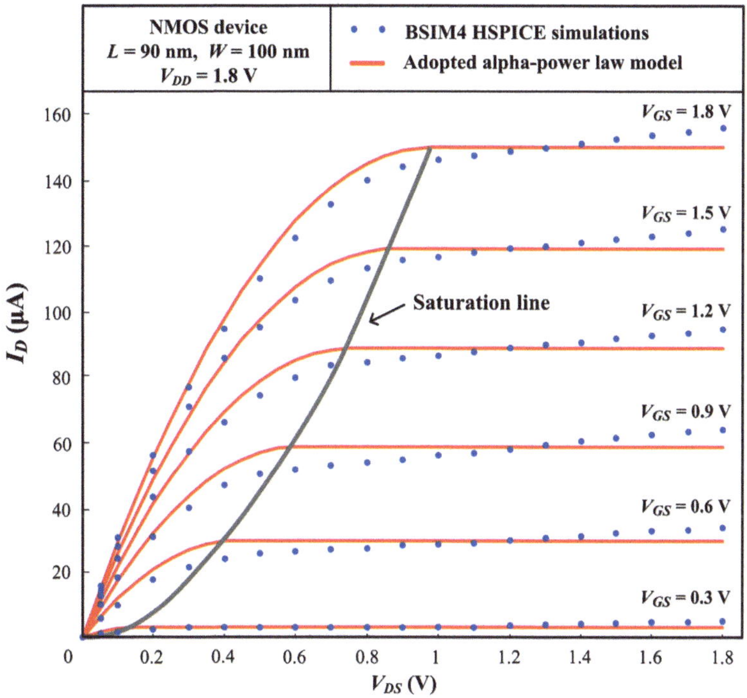

Fig. 5.3 The 90 nm NMOS device I–V characteristics

For the device drain current, the following accurate and simple version of the alpha-power law MOSFET model [17, 18] is used.

In triode or linear region ($V_{DS} \leq V'_{DO}$),

$$I_D = B(V_{GS} - V_T)^{\alpha}\left(2 - \frac{V_{DS}}{V'_{DO}}\right)\frac{V_{DS}}{V'_{DO}} \tag{5.4}$$

In saturation region ($V_{DS} > V'_{DO}$),

$$I_D = B(V_{GS} - V_T)^{\alpha}, \tag{5.5}$$

$$V'_{DO} = K(V_{GS} - V_T)^{\alpha/2} \tag{5.6}$$

The parameters of the adopted MOSFET model are determined by the used technology process or extracted by suitable fitting of device characteristics produced by simulation using BSIM4 predictive models [19, 20]. α is the velocity saturation index, V'_{DO} is the drain-source saturation voltage, B is the transconductance parameter extracted for $V_{GS} = V_{DD}$, and $V_{DS} = 3/4 \cdot V_{DD}$ (NMOS device), $V_{DS} = 4/5 \cdot V_{DD}$ (PMOS device), V_T is the threshold voltage of the device, and K is a constant determined by the drain-source saturation voltage at $V_{GS} = V_{DD}$. Such drain current equations model with sufficient accuracy the behavior of near − 100 nm devices (Fig. 5.3).

Fig. 5.4 Initial operation regions of the primitive circuit

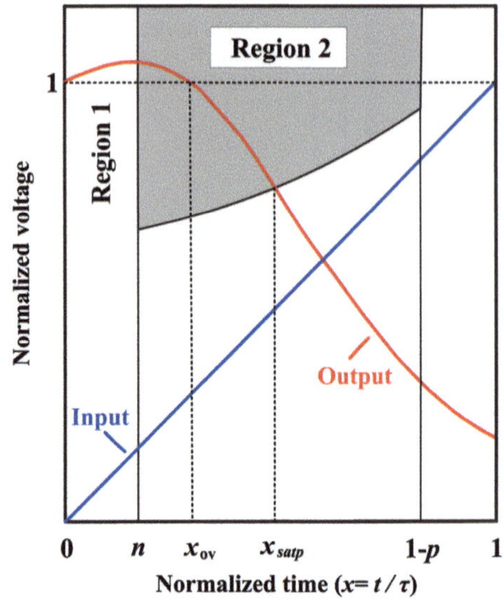

The PMOS device current in the triode region (i.e., for $1 - u_{\text{out}} < u'_{\text{dop}}$) is given by the following equation:

$$I_{\text{p}} = \text{k}_{\text{lp1}}\,(1 - x - p)^{\alpha_{\text{p}}/2}(1 - u_{\text{out}}) - \text{k}_{\text{lp2}}(1 - u_{\text{out}})^2, \tag{5.7}$$

where p is the normalized threshold voltage of the device and k_{lp1}, k_{lp2} are constants dependent on the supply voltage and on the parameters K, B, and α of the device.

The NMOS device current in the saturation region (i.e., for $u_{\text{out}} \geq u'_{\text{don}}$) is given by

$$I_{\text{n}} = \text{k}_{\text{sn}}(x - n)^{\alpha_{\text{n}}} \tag{5.8}$$

where n is the normalized threshold voltage of the device and k_{sn} is a constant dependent on the supply voltage and on the parameters B and α of the device.

For the evaluation of the short-circuit energy dissipation, analytical expressions of the output waveforms in the two first inverter operating regions (Fig. 5.4) are required. For $0 \leq x \leq n$, the NMOS device is off and the PMOS device is in the triode region whereas for $n < x \leq x_{\text{satp}}$, the NMOS device is saturated and the PMOS device remains in the triode region. Part of the charge from the input, injected through the gate-to-drain coupling capacitance, causes an overshoot at the early part of the output voltage waveform. During the overshoot, there is no current from power supply to ground because the output voltage is greater than the supply voltage. In region 1, the differential Eq. (5.2) cannot be solved analytically. The main influence on the output voltage waveform in this region is due to the charge from the input, injected through the gate-to-drain coupling capacitance, whereas the influence of the PMOS transistor current is quite smaller, so if one applies careful approximations in the expression of the PMOS device current, the accuracy of the resulting output

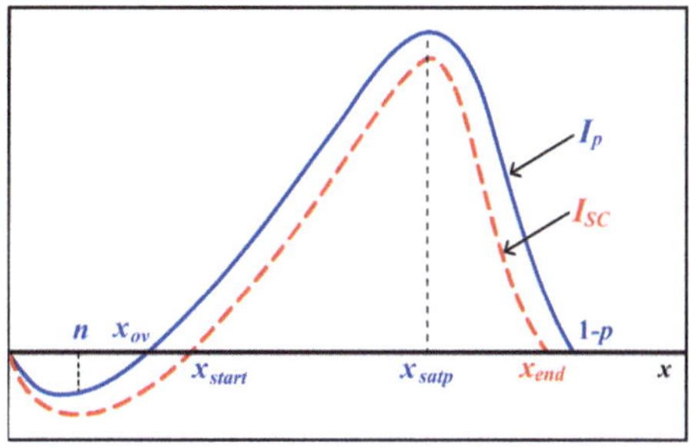

Fig. 5.5 PMOS device current and short-circuit current waveforms

voltage expression will not be affected. Therefore, an average value of x ($x = n/2$) is used in the first term of PMOS current (Eq. (5.7)), and an approximated expression for u_{out} ($u_{out} = 1 + c_m \cdot x$) is used in the quadratic term of the PMOS current. This approximated expression takes into account only the charge through the gate-to-drain coupling capacitance, but it is used only in the quadratic term of the PMOS transistor current [i.e., $(1 - u_{out})^2$]. The approximation is reasonable since in this operating region, the main influence on the output voltage waveform is due to the charge injected through the gate-to-drain coupling capacitance. Moreover, the charge contributed by the quadratic term of the PMOS transistor current is very small due to the small values of the PMOS device normalized drain-source voltage (i.e., $1 - u_{out}$) in this operating region. After the adoption of both approximations, u_{out} is given as

$$u_{out} = 1 + \frac{c_m}{C^3 A_{lp1}^3}[2 A_{lp2} c_m (e^{-x C A_{lp1}} - 1) + 2 C A_{lp1} A_{lp2} c_m x +$$

$$C^2 A_{lp1}^2 (1 - e^{-x C A_{lp1}} - A_{lp2} c_m x^2)], \quad (5.9)$$

where A_{lp1}, A_{lp2}, and C are constants dependent on the supply voltage, the normalized threshold voltages n and p, the load and gate-drain capacitances, the input transition time, and the parameters K, B, and α of the PMOS device.

In order to obtain an expression of the output voltage waveform for $n < x \le x_{satp}$, the PMOS current (Fig. 5.5) is approximated by a linear function of x:

$$I_p = I_{pn} + S_1(x - n), \quad (5.10)$$

where I_{pn} is the PMOS device current for $x = n$. The current slope S_1 is computed by equating the exact PMOS current in the triode region with the approximated one at $x = (1 - p)/2$.

By using the linear approximation of the PMOS current, the differential Eq. (5.2) in region 2 is solved and the output voltage waveform is described by

$$u_{out} = u_n + (x - n)c_m + dI_{pn}(x - n) + \frac{dS_1(x - n)^2}{2} - \frac{A_{sn}(x - n)^{\alpha_n + 1}}{\alpha_n + 1}, \quad (5.11)$$

where A_{sn} and d are constants dependent on the supply voltage, the load and gate-drain capacitances, the input transition time, and the parameters B and α of the NMOS device, and u_n is the normalized output voltage at $x = n$.

The short-circuit energy dissipation for a rising input is the energy of the current (I_{SC}), which is provided from the power supply (Fig. 5.2). The current through the PMOS device includes two non-short-circuit current components: the current flowing through C_{gsp} and the current flowing from the output to the supply node during the overshoot of the output signal. The short-circuit energy dissipation during the falling output transition is defined as

$$E_{SC} = V_{DD} \int_{x_{start}}^{x_{end}} I_{SC}\, \tau \mathrm{d}x = V_{DD} \left(\int_{x_{start}}^{x_{satp}} I_{SC}\, \tau \mathrm{d}x + \int_{x_{satp}}^{x_{end}} I_{SC}\, \tau \mathrm{d}x \right), \tag{5.12}$$

where $I_{SC} = I_p - I_{C_{gsp}}$ and $I_{C_{gsp}} = C_{gsp}(V_{DD}/\tau)$.

In the first integral of (5.12), the following linear approximation of the PMOS device current is used:

$$I_p = S_2(x - x_{ov}). \tag{5.13}$$

The current slope S_2 is computed by equating the exact PMOS current in the triode region with that of (5.12), at the middle of the interval $[x_{ov}, x_{satp}]$. In the second integral of (5.12), the exact PMOS saturation current expression is used. After that, the inverter short-circuit energy dissipation is computed as

$$E_{SC} = \frac{V_{DD}}{2}(x_{satp} - x_{start}) \left[(x_{satp} + x_{start} - 2x_{ov})S_2 - \frac{2C_{gsp}V_{DD}}{\tau} \right] +$$
$$\frac{V_{DD}k_{sp}\tau}{\alpha_p + 1}[(1 - p - x_{satp})^{\alpha_p+1} - (1 - p - x_{end})^{\alpha_p+1}] - C_{gsp}V_{DD}^2(x_{end} - x_{satp}). \tag{5.14}$$

The computation of the boundary values x_{start} and x_{end} is achieved by setting the short-circuit current (I_{SC}) to zero, when the PMOS device operates in the linear and saturation region, respectively:

$$S_2(x_{start} - x_{ov}) - C_{gsp}(V_{DD}/\tau) = 0, \tag{5.15}$$

$$k_{sp}(1 - x_{end} - p)^{\alpha_p} - C_{gsp}(V_{DD}/\tau) = 0. \tag{5.16}$$

For the computation of the normalized time point x_{satp}, suitable second-order Taylor series expansions of u_{out} and u'_{dop} should be used. In Fig. 5.6a, the inverter short-circuit energy per transition is plotted as a function of the input transition time and for two different values of output load. In addition, in Fig. 5.6b, the inverter short-circuit energy per transition is plotted as a function of the supply voltage. The channel

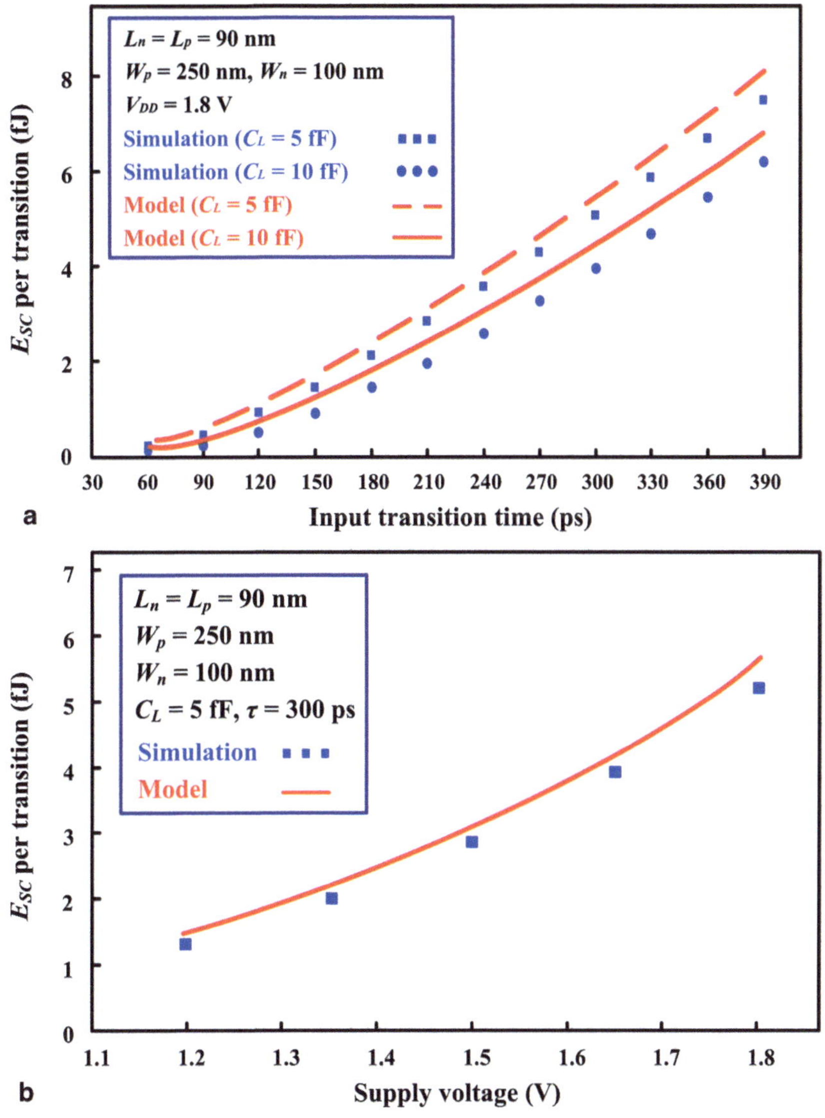

Fig. 5.6 **a** Inverter short-circuit energy dissipation per input transition for two different values of output load, **b** Inverter short-circuit energy dissipation per supply voltage

length of the used devices was 90 nm, and the results show very good agreement with BSIM4 [21] and HSPICE [22] simulations.

As described in the case of short-circuit energy dissipation modeling, the combination of saturation, triode, and cutoff devices' operation modes defines different operating regions of the CMOS inverter. For the separation of the inverter operation into regions, the input voltage slope is also considered. Such definition of operating regions can be extended to the overall inverter operation range as shown in Fig. 5.7 [3].

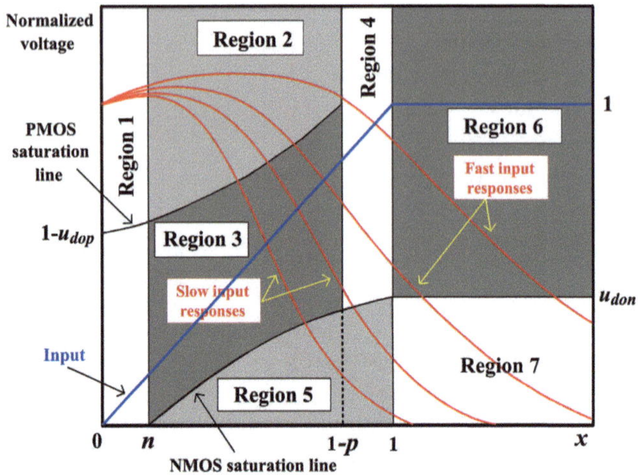

Fig. 5.7 Operating regions of the primitive circuit [3]

The differential equation can be defined and solved for all operating regions by using appropriate approximations when needed [3]. Then, the analytical output waveform expressions can be used for the derivation of explicit formulae for the computation of the primitive circuit's dynamic characteristics, such as propagation delay and output voltage transition time.

The inverter model can be extended to more complex CMOS primitive circuits such as multi-input CMOS gates [6, 13, 14]. Such extension is performed by reducing each gate to an equivalent inverter. This procedure requires the modeling of the series-connected and parallel-connected transistors (i.e., their reduction to single equivalent transistors), and the reduction of overlapping inputs to a single effective input signal. For a successful reduction of a gate, one should take into account the transition time of the gate inputs, the number of switching inputs of the gate, the position of the switching inputs, the body effect, the output load, and the internal node capacitances.

5.3 Compact Modeling of MOSFET Devices

The accuracy and computational efficiency of primitive circuit analysis models is directly affected by the accuracy and the simplicity of the MOSFET model used. Such a model must meet two equally important requirements: (1) accurate transistor drain current prediction and (2) simplicity of the device model to obtain explicit expressions for design parameters (i.e., transient response and energy dissipation).

In most existing current device models, the effects that determine the device behavior are accounted for through physical and empirical parameters. With the growing complexity of physical mechanisms in nanometer devices, device models become very complex and employ a large number of parameters to provide the highest

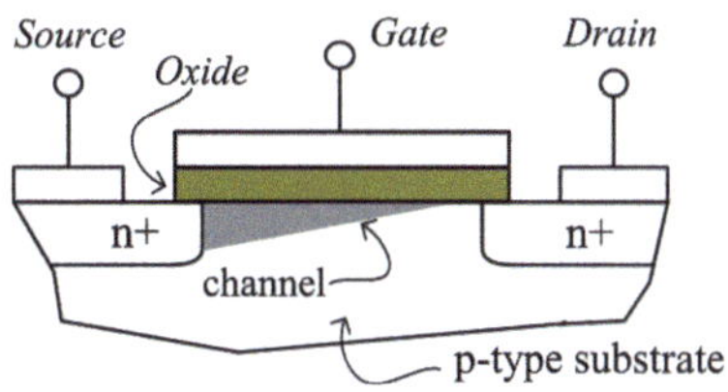

Fig. 5.8 The MOSFET transistor

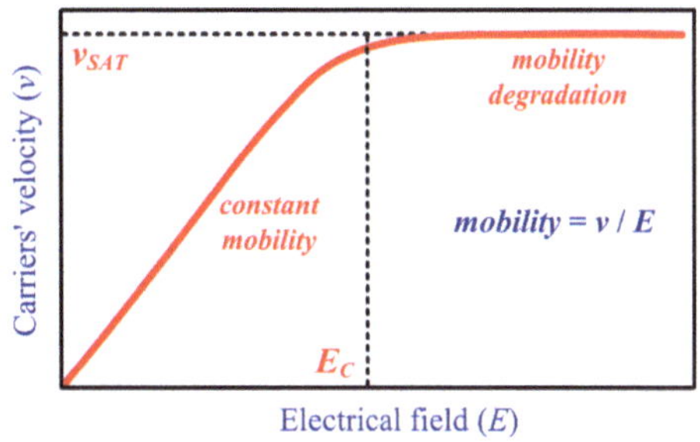

Fig. 5.9 Electric field to carrier velocity dependence

accuracy [11, 12]. Although these complex but accurate models can be handled by circuit simulators, they do not satisfy the requirement of computational efficiency. Hence, compact device models are needed, as simple as possible, to take into account the influences of essential physical mechanisms in nanometer devices by using few parameters extracted through measurements or simulations [12].

In order to strengthen the modeling methodology of CMOS circuit primitives, an accurate and compact I–V model for nanometer MOSFETs is required. The I–V equations of such model may use empirical parameters to match measured or simulated device characteristics, and should take into account the influence of predominant effects in nanometer devices, such as [11, 12, 23]: mobility degradation and velocity saturation, channel-length modulation, drain-induced barrier lowering (DIBL), body effect, narrow-channel width effect, and source-drain parasitic resistance.

5.3.1 *Modeling of Small Dimension Effects in MOSFET devices*

In MOSFET devices (Fig. 5.8), for small electric fields, the carriers' mobility is constant and independent of the applied electric field. As shown in Fig. 5.9, when the horizontal electrical field (moving the channel carriers) reaches a critical value, the carriers' velocity tends to saturate owing to scattering effect (i.e., electrons moving in semiconductor material collide with silicon atoms). This effect is more pronounced for reduced channel length that implies higher horizontal electric fields for equivalent drain-source voltages [23]. The vertical electric field originating from the gate voltage further inhibits channel carrier mobility. This field pushes carriers toward the gate oxide and the carriers' mobility is reduced due to carrier collisions with the oxide–channel interface.

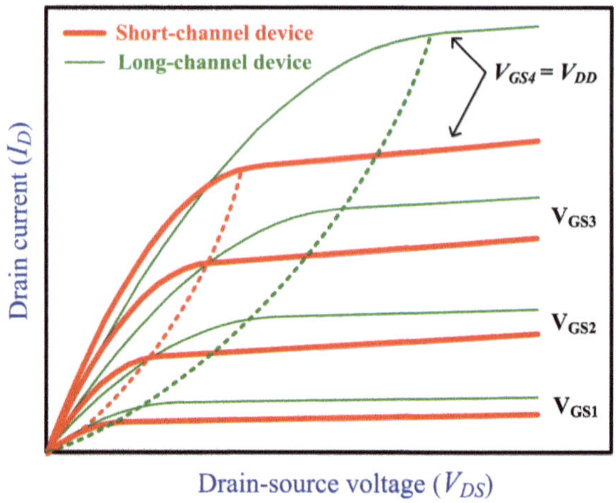

Fig. 5.10 MOSFET characteristics in the absence and presence of velocity saturation

The influence of mobility degradation and velocity saturation effects on the MOSFET device output characteristic curves is illustrated in Fig. 5.10. In short-channel devices, the saturation occurs at smaller drain-source voltages, and the spacing of the I–V curves in saturation is not according to square law, but becomes nearly proportional to gate-source voltage increment. As described in [17] as well as by (5.5) and (5.6), the modeling of mobility degradation and velocity saturation effects is achieved by employing the velocity saturation index (α) to describe the power laws featuring the drain current and the drain-source saturation voltage.

Channel length modulation (CLM) refers to the shortening of the length of the inverted channel region with increase in drain bias. When the device operates in saturation and the drain voltage increases, the uninverted region at the vicinity of the drain (pinch-off region) expands toward the source, shortening the length of the channel region (Fig. 5.8). Due to the fact that resistance is proportional to length, shortening the channel decreases its resistance, causing an increase in current with increase in drain bias for a device operating in saturation [1, 11]. The effect is more pronounced when the source-to-drain separation is short (i.e., in deep-submicrometer and nanometer devices) [23].

In a MOSFET device, a potential barrier exists between the source and the channel, which is controlled by the gate voltage. When the gate voltage is increased, the barrier between the source and the channel is decreased, increasing the carriers' injection from the source to the channel over the lowered barrier. In very short-channel devices, as drain voltage increases, more depletion is performed by the drain bias, and the electric field at the drain penetrates to the source region causing an additional decrease of the barrier at source (Fig. 5.11). This is referred to as DIBL effect [11]. As a result, the device can conduct significant drain current due to an increase in carriers injected from the source. DIBL affects the drain current versus drain bias curve, causing the current to increase with drain bias in the saturation region of operation (i.e., at high drain-to-source voltages). This current increase is additional to that caused by the CLM effect. The dependence between the drain current and the drain-source voltage

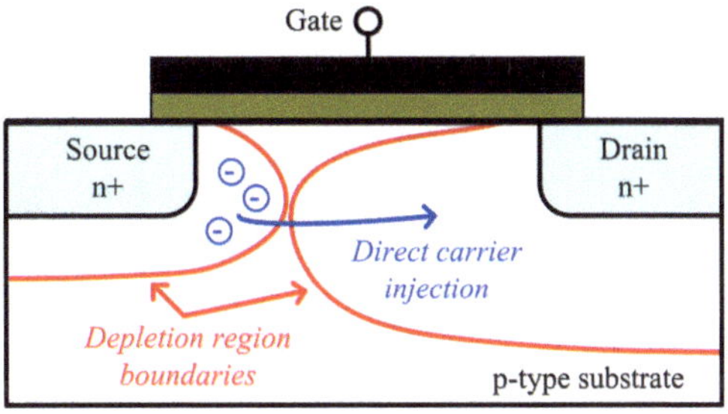

Fig. 5.11 Drain-induced barrier lowering (DIBL) effect in a MOSFET device

in the saturation region, which is due to CLM and DIBL effects, could be modeled through the inclusion of two additional empirical fitting parameters (A, D):

$$I_D = B(V_{GS} - V_T)^{\alpha}[A + D(V_{DS} - V_{DO})], \tag{5.17}$$

where V_{DO} is the drain-source saturation voltage of the device at $V_{GS} = V_{DD}$. Given the practical target of the model, that is, the analytical modeling of primitive circuits' operations, this linear dependence maintains the simplicity, while providing the required accuracy by including the influence of both effects.

When a positive source-bulk voltage (V_{SB}) is applied, the bulk is at a negative potential with respect to the source, and this increases the depletion between the source and the bulk. The minority electrons attracted from the p-type bulk have to overcome this increase in depletion, and therefore the gate voltage required to form and maintain an inversion layer or channel (i.e., threshold voltage) becomes higher. This is referred to as body effect [11, 23]. For the determination of the device threshold voltage when V_{SB} is positive, a linear approximation (5.18) of the BSIM4 model [21] expression (5.19), describing the body effect, is used

$$V_{TH} = V_{TO} + \gamma V_{SB}, \tag{5.18}$$

$$V_{TH} = V_{TO} + K_1(\sqrt{\varphi_s + V_{SB}} - \sqrt{\varphi_s}) + K_2 V_{SB}. \tag{5.19}$$

where V_{TO} is the threshold voltage for $V_{SB} = 0$, φ_s is the inversion surface potential, K_1, K_2 are the BSIM4 body effect coefficients, and γ is the simplified body effect coefficient.

In MOSFET devices fabricated by using the local oxidation of silicon (LOCOS) process (Fig. 5.12) [23], the depletion region is not limited to just the area below the thin oxide, since the polysilicon gate overlaps the field oxide on both sides of the channel region, along the width direction of the device. For large device widths, the part of the depletion region on the sides is a small percentage of the total depletion region. As the device width is scaled down, the depletion charge under the gate is reduced but the fringing charge remains relatively unchanged, constituting a significant proportion. The gate is responsible for depleting a larger region and hence, higher gate voltage is required, resulting in increased threshold voltage. In effect, this results in lower driving capability per width unit of narrow-width devices in comparison with the driving capability per width unit of wide-width devices. The

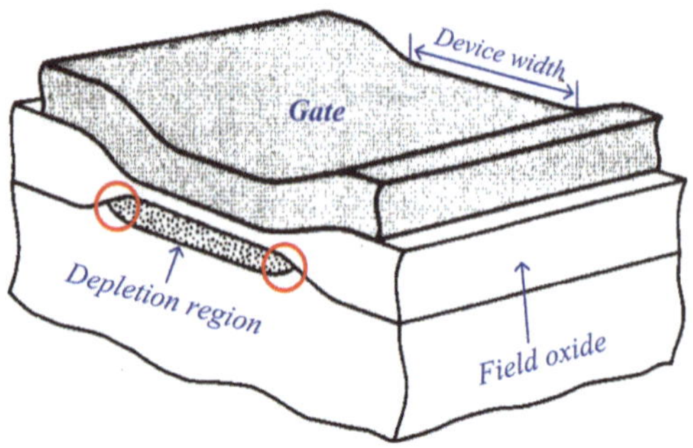

Fig. 5.12 Cross-section along the width of a MOSFET device in which the polysilicon gate overlaps the field oxide of the device [23]

prediction of the drain current for varying device widths (i.e., the inclusion of narrow-channel width effects) is obtained by computing the transconductance parameter B of the device as a quadratic function of the device channel width (W) [24]:

$$B = \beta_1 + \beta_2 W + \beta_3 W^2, \tag{5.20}$$

where the coefficients β_i are determined by fitting the quadratic plot to the B versus W plot, once for a given nanometer technology.

It has to be mentioned that for devices fabricated with the shallow-trench isolation process [23], the fringing field from the gate regions beyond the channel edges support depletion charges in the channel, and in contrast to the LOCOS process, the fringing filed makes the depletion region deeper, thus increasing the surface potential, and helping the start of the inversion layer. In effect, this results in higher driving capability per width unit of narrow-width devices in comparison with that of wide-width devices. However, the transconductance parameter of the device can be predicted by using the aforementioned technique that is based on (5.20).

In long-channel devices, the source-drain parasitic resistance is negligible compared with the channel resistance. However, in very short-channel devices, it can be an appreciable fraction of the channel resistance and can therefore cause significant current degradation [11]. The most severe current degradation occurs in the triode region, i.e., for low values of drain-source voltage. This is because the channel resistance is low (the slope of the drain current versus drain-source voltage curve is high) under such bias conditions. An expression-based explanation can be easily derived by Eqs. (5.21)–(5.23).

$$I_{\mathrm{D}} = \frac{V_{\mathrm{DS}}}{R_{\mathrm{ch}} + R_{\mathrm{sd}}}, \tag{5.21}$$

$$R_{\mathrm{ch}} = \frac{V_{\mathrm{DS}}}{I_{\mathrm{D-without}\ R_{\mathrm{sd}}}}, \tag{5.22}$$

$$I_{\mathrm{D}} = \frac{I_{\mathrm{D-without}\ R_{\mathrm{sd}}}}{1 + (R_{\mathrm{sd}} I_{\mathrm{D-without}\ R_{\mathrm{sd}}})/V_{\mathrm{DS}}}. \tag{5.23}$$

Since the drain current dependence of the drain-source voltage is small in saturation, the current in this region is least affected by the parasitic resistance. The source-drain parasitic resistance effect can be taken into account by adopting lower transconductance parameter in triode region than that of the saturation region. The transconductance parameters in both operating regions are derived by using proper fitting points on the device I–V output characteristics, as described in Sect. 5.3.2.

5.3.2 *Putting it All Together: A Compact MOSFET Current Model*

After the discussion of the predominant effects that impact the transistor behavior when technology scales down, compact model equations are introduced to describe the drain current of nanometer devices. Such equations take into account effects such as: mobility degradation and velocity saturation, channel-length modulation, DIBL, body effect, narrow-channel width effect, and source-drain parasitic resistance.

As mentioned earlier in this chapter, the objective of the derived device current equations is to strengthen the modeling methodology of CMOS circuit primitives. The adopted approach is based on the derivation of compact expressions that fit transistor curves over both bias ranges. The expressions use empirical parameters to specifically match measured device characteristics [25]. Empirically based models afford the possibility of developing analytical timing and energy dissipation models for CMOS primitive circuits, without the necessity of expensive numerical iterations [7, 25].

The combination of the advantages of existing compact models [14, 17, 18] with the modeling of the effects mentioned in Sect. 5.3.1, leads to the following strong inversion drain current MOSFET model:

For $V_{DS} \leq V'_{DO}$ (triode region),

$$I_D = B_{tri}(V_{GS} - V_{TH})^{\alpha}\left(2 - \frac{V_{DS}}{V'_{DO}}\right)\frac{V_{DS}}{V'_{DO}}. \tag{5.24}$$

For $V_{DS} > V'_{DO}$ (saturation region),

$$I_D = B_{sat}(V_{GS} - V_T)^{\alpha}[\mathrm{A} + \mathrm{D}(V_{DS} - V_{DO})]. \tag{5.25}$$

where

$$V'_{DO} = V_{DO}\left(\frac{V_{GS} - V_T}{V_{DD} - V_T}\right)^{\frac{\alpha}{2}}, \tag{5.26}$$

$$V_T = V_{TO} + \gamma V_{SB}, \quad B_{tri} = \frac{I'_{DO}}{(V_{DD} - V_T)^{\alpha}}, \quad B_{sat} = \frac{I_{DO}}{(V_{DD} - V_T)^{\alpha}},$$

$$\mathrm{A} = \frac{I'_{DO}}{I_{DO}}, \text{ and } \mathrm{D} = \frac{1 - \mathrm{A}}{V_{DD} - V_{DO}}.$$

V_{TO} stands for the zero back-gate bias threshold voltage, γ is the coefficient accounts for the body effect, V_{DO} is the drain-source saturation voltage at $V_{GS} = V_{DD}$, and

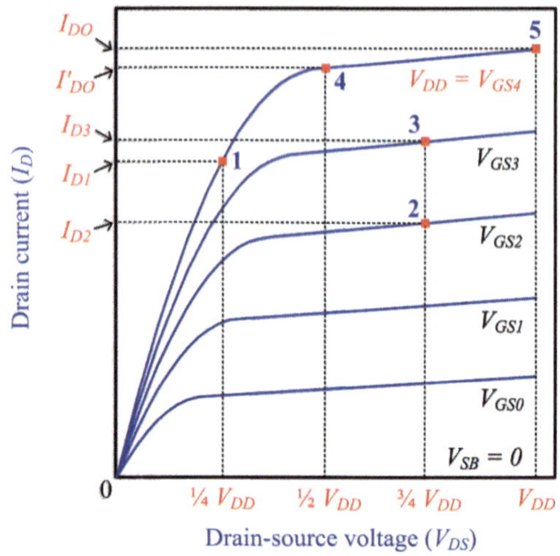

Fig. 5.13 Selected fitting points for extracting parameters of the NMOS device (the fitting points for the PMOS device are selected similarly according to the mentioned details) [25]

$B_{\rm tri}$ and $B_{\rm sat}$ are transconductance parameters for triode and saturation region, respectively. $I_{\rm DO}$ is the drain current at $V_{\rm GS} = V_{\rm DS} = V_{\rm DD}$, whereas $I'_{\rm DO}$ is the drain current at $V_{\rm GS} = V_{\rm DD}$, $V_{\rm DS} = 1/2 \cdot V_{\rm DD}$ (for the NMOS device), and $V_{\rm DS} = 2/3 \cdot V_{\rm DD}$ (for the PMOS device).

The model parameters are extracted by using appropriately selected fitting points on the device I–V curves, as illustrated in Fig. 5.13 [25]. $I_{\rm DO}$ and $I'_{\rm DO}$ are easily obtained from the output MOSFET characteristics (points 5 and 4, respectively). α is extracted from the following equation, which is derived from (5.25) by using the fitting points 2 and 3 [17]:

$$\alpha = \frac{\ln\left(\frac{I_{\rm D3}}{I_{\rm D2}}\right)}{\ln\left(\frac{V_{\rm GS3}-V_{\rm TO}}{V_{\rm GS2}-V_{\rm TO}}\right)}. \tag{5.27}$$

$V_{\rm DO}$ is computed by combining (5.24) and (5.26) and using the fitting point 1:

$$V_{\rm DO} = \frac{I'_{\rm DO}V_{\rm DD} + V_{\rm DD}\sqrt{I'_{\rm DO}(I'_{\rm DO} - I_{\rm D1})}}{4I_{\rm D1}}. \tag{5.28}$$

By simulating the device, we obtain $I_{\rm DO}$ and $I'_{\rm DO}$ and consequently $B_{\rm tri}$ and $B_{\rm sat}$ for the minimum and the maximum used device width, as well as for few intermediate width values, in order to fit $B_{\rm tri}$ versus W and $B_{\rm sat}$ versus W plots to quadratic plots, such as:

$$B_{\rm tri} = \beta_{\rm t1} + \beta_{\rm t2}W + \beta_{\rm t3}W^2, \tag{5.29}$$

$$B_{\rm sat} = \beta_{\rm s1} + \beta_{\rm s2}W + \beta_{\rm s3}W^2. \tag{5.30}$$

Given the applicative target of the presented MOSFET current model that is the analysis and modeling of primitive CMOS circuits, an accurate characterization

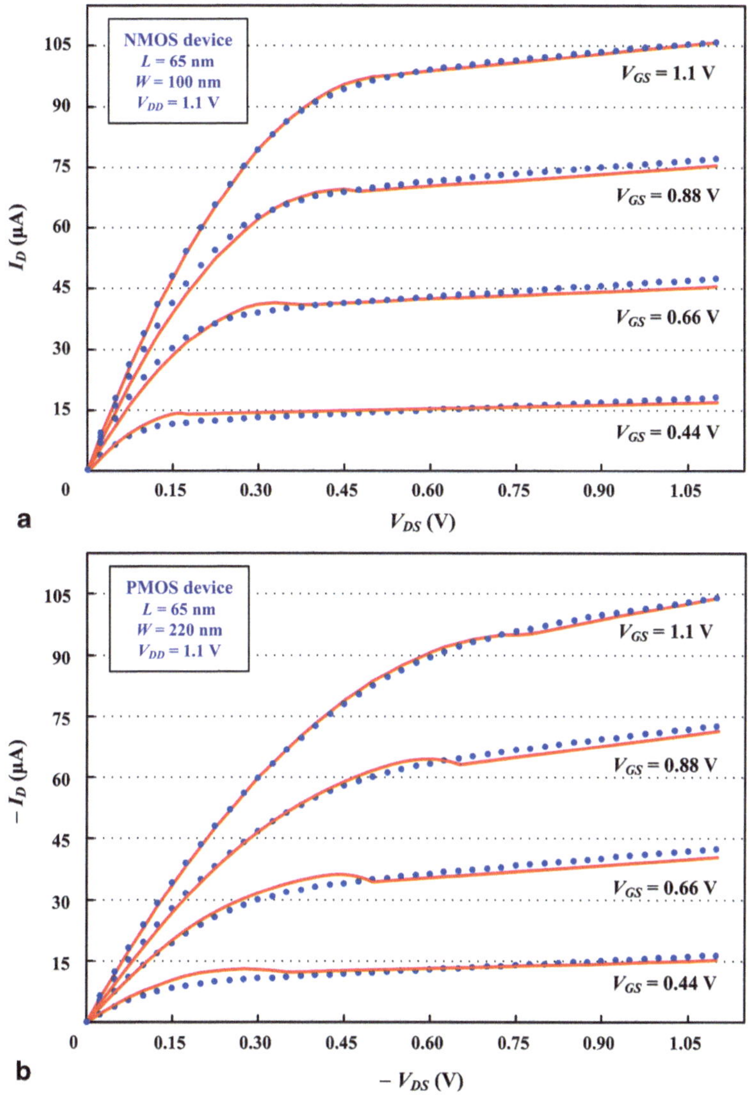

Fig. 5.14 I–V plots for 65 nm CMOS technology, **a** NMOS device, **b** PMOS device (*continuous lines* correspond to model results and *dots* correspond to BSIM4 HSPICE simulations)

of the current behavior in the subthreshold operating region is not essential. The presented MOSFET current model can lead to accurate estimation of the dynamic characteristics of CMOS primitive circuits, even considering a current abruptly going to zero for $V_{GS} = V_T$, and not modeling second-order effects (such as DIBL) within the threshold voltage expression.

The derived MOSFET current model has been validated by using a 65 nm CMOS technology. Figure 5.14 presents the NMOS and PMOS output characteristic curves for device widths of 100 nm and 220 nm, respectively. In Fig. 5.15, the

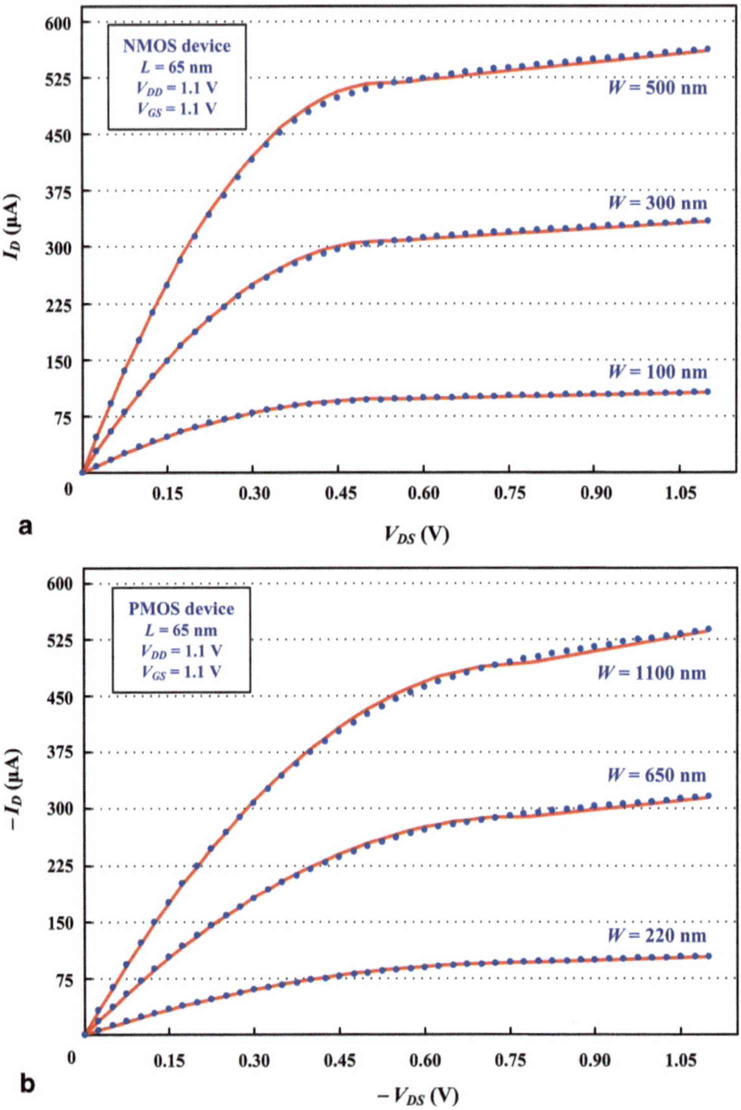

Fig. 5.15 I–V plots for different device channel widths, **a** NMOS device, **b** PMOS device (*continuous lines* correspond to model results and *dots* correspond to BSIM4 HSPICE simulations)

model is validated for several device widths. Continuous lines correspond to the model results, whereas dots correspond to the BSIM4 HSPICE simulations. The experimental results derived by the model show very good agreement with the simulation data extracted by using predictive technology models [19, 20].

The exhibited discontinuity at the boundary between triode and saturation operating regions is due to the fact that a basic concern of the device current model is to avoid complex dependence on the drain-source voltage. Such discontinuity could cause problems in circuit simulators that require continuity of the functions, but should not cause problems in case of use for the analytical computation of dynamic characteristics in primitive circuits.

5.4 Conclusion

In this chapter, a methodology for the modeling of dynamic characteristics (such as transient response and energy dissipation) of CMOS primitive circuits has been presented. As a case study, the CMOS inverter has been used. Following a detailed analysis of its operation, accurate formulae for its output response are derived for the different operating regions, and based on this analysis, analytical expressions for the calculation of the timing and energy parameters can be produced. The derived models account for the influences of circuit design and operational parameters, as well as of device parameters. After an analysis of predominant effects in modern nanometer device technologies, a compact device current model has been presented that exhibits simplicity and is accurate enough to strengthen the modeling methodology of CMOS circuit primitives.

References

1. Weste, N. H. E., Harris, D. M. (2011). *CMOS VLSI design: A circuits and systems perspective*. Boston: Pearson Education.
2. Rabaey, J. (2009). *Low power design essentials*. New York: Springer.
3. Bisdounis, L., Nikolaidis, S., Koufopavlou, O. (1998). Analytical transient response and propagation delay evaluation of the CMOS inverter for short-channel devices. *IEEE Journal of Solid-State Circuits, 33*(2), 302–306.
4. Rossello, J. L., Segura, J. (2004). An analytical charge-based compact delay model for submicrometer CMOS inverters. *IEEE Transactions on Circuits and Systems I, 51*(7), 1301–1311.
5. Kabbani, A., Al-Khalili, D., Al-Khalili, A. J. (2003). Technology-portable analytical model for DSM CMOS inverter transition time estimation. *IEEE Transactions on Computer-Aided Design of Integrated Circuits and Systems, 22*(9), 1177–1187.
6. Bisdounis, L., Koufopavlou, O. (2000). Short-circuit energy dissipation modeling for submicrometer CMOS gates. *IEEE Transactions on Circuits and Systems I, 47*(9), 1350–1361.
7. Consoli, E., Giustolisi, G., Palumbo, G. (2012). An accurate ultra-compact I-V model for nanometer MOS transistors with applications on digital circuits. *IEEE Transactions on Circuits and Systems I, 59*(1), 159–169.
8. Shih, Y. H., Leblebici, Y., Kang, S. M. (1993). ILLIADS: A fast timing and reliability simulator for digital MOS circuits. *IEEE Transactions on Computer-Aided Design of Integrated Circuits and Systems, 12*(9), 1387–1402.
9. Kutuk, H., Goknar, I. C., Kang, S. M. (1999). Interconnect simulation in a fast timing simulator ILLIADS-I. *IEEE Transactions on Circuits and Systems I, 46*(1), 178–189.
10. Bellido, M. J., Juan, J., Valencia, M. (2006). *Logic-timing simulation and the degradation delay model*. London: Imperial College Press.

11. Taur, Y., Ning, T. (2009). *Fundamentals of modern VLSI devices*. Cambridge: Cambridge University Press.
12. Bhattacharyya, A. B. (2009). *Compact MOSFET models for VLSI design*. Singapore: Wiley.
13. Chatzigeorgiou, A., Nikolaidis, S., Tsoukalas, I. (1999). A modeling technique for CMOS gates. *IEEE Transactions on Computer-Aided Design of Integrated Circuits and Systems, 18*(5), 557–575.
14. Sakurai, T., Newton, A. R. (1991). Delay analysis of series-connected MOSFET circuits. *IEEE Journal of Solid-State Circuits, 26*(2), 122–131.
15. Liu, D., Svensson, C. (1994). Power consumption estimation in CMOS VLSI chips. *IEEE Journal of Solid-State Circuits, 29*(6), 663–670.
16. Bisdounis, L. (2010). Short-circuit energy dissipation model for sub –100 nm CMOS buffers. In Proceedings of the 17th IEEE international conference on electronics, circuits and systems, Athens, 12–15 December 2010, pp. 615–618.
17. Sakurai, T., Newton, A. R. (1990). Alpha-power law MOSFET model and its applications to CMOS inverter delay and other formulas. *IEEE Journal of Solid-State Circuits, 25*(2), 584–594.
18. Nose, K., Sakurai, T. (2000). Analysis and future trend for short-circuit power. *IEEE Transactions on Computer-Aided Design of Integrated Circuits and Systems, 19*(9), 1023–1030.
19. Zhao, W., Cao, Y. (2007). Predictive technology model for nano-CMOS design exploration. *ACM Journal of Emerging Technologies in Computing Systems, 3*(1), 2–17.
20. Arizona State University (2006). Predictive technology model. http://ptm.asu.edu/cgi-bin/test/nanocmos.cgi.
21. Liu, W. (2001). *MOSFET models for SPICE simulation including BSIM 3v3 and BSIM4*. New York: Wiley.
22. Synopsys Inc. (2007). *HSPICE simulation and analysis user guide*. Mountain View: Synopsys Inc.
23. Tsividis, Y. (2003). *Operation and modeling of the MOS transistor*. New York: Oxford University Press.
24. Chandra, N., Kumar, A., Bhattacharyya, A. B. (2009). Extended Sakurai-Newton MOSFET model for ultra-deep-sub-micrometer CMOS digital design. In Proceedings of the 22nd international conference on VLSI design, New Delhi, 5–9 January 2009, pp. 247–252.
25. Bisdounis, L. (2012). An accurate and compact MOSFET I-V model for nanometer CMOS circuit analysis. In Proceedings of the 2nd pan-hellenic conference on electronics and telecommunications, Thessaloniki, 16–18 March 2012.

Chapter 6
From Hardware Security Tokens to Trusted Computing and Trusted Systems

Apostolos P. Fournaris and Georgios Keramidas

Abstract As security attacks are becoming an everyday real-life scenario, security engineers must invent more intricate countermeasures to deal with them. Infusion of strong security to a computer system by recruiting specialized hardware tokens has already an established foothold in the modern information technology (IT) world. However, these tokens nowadays must be appropriately adapted to ensure not only strong security but also trust. Modern security specialists believe that the ultimate security goal is not only to provide a strong security shield but also to guarantee in an undeniable way that a system is trusted (the system always performs its intended functionality).

In this chapter, we elaborate on the IT world's transition from security to trust and describe the trusted computing approach to provide trusted systems. Current trends are presented and tools like trusted virtualization approaches are analyzed. The trusted computing technology features are discussed and our critical view on what the trusted computing future will be like is offered.

6.1 Introduction

Information technology (IT) applications whether on smart phones, tablets, desktop, notebooks, netbooks, PCs have penetrated in our everyday life and handle a large number of our daily data transactions including personal, business, social, or governmental type of communications. A great deal of sensitive data are involved in those transactions and if these data fall in the wrong hand they can cause critical security damage. In parallel to this evolution, security threats have also become smarter, leading into significant risks. Such threats include [1]:

A. P. Fournaris (✉)
Electrical and Computer Engineering Department, University of Patras, Greece.
KNOSSOSnet Research Group, Informatics and MM Department,
Technological Educational Institute of Patras, Greece
e-mail: apofour@ieee.org

G. Keramidas
Electrical and Computer Engineering Department, University of Patras, Patras, Greece
e-mail: keramidas@ece.upatras.gr

N. Sklavos et al. (eds.), *System-Level Design Methodologies for Telecommunication*,
DOI 10.1007/978-3-319-00663-5_6,

- Vulnerable programs (coding bugs, buffer overflows, parsing errors)
- Malicious programs (spyware, Trojans)
- Misconfigured programs (security features not turned on)
- Social engineering (phishing/pharming attacks)
- Physical theft (laptops, smart phones)
- Electronic eavesdropping (capturing email)

Most of above threads are completely transparent to the computer system user as well as other associated systems (e.g., operators). Every transaction with a computer system can potentially lead to unauthorized security data leakage. In other words, during operation, the computer system is considered untrusted. The associated risk with such system usage is not always affordable, especially when it comes to transactions involving sensitive data. In such cases, the user must be confident that his system can be trusted. As a result, trust is a very important and desirable feature of modern computer systems and considerable effort has been invested in the development of trusted systems.

In this chapter, we offer a brief overview of the approaches followed over time to protect a computing system from security threads, starting from simple tokens, e.g., magnetic cards, and moving toward the design of a fully trusted system. Since software can always be tampered and manipulated, the realization of a trusted system can only be achieved through dedicated hardware components that offer untampered, secure areas, an attacker cannot reach or violate. So, we base our analysis on how hardware structures can be utilized to provide trust. Moreover, the notion of trust is thoroughly defined and trust models that lead to trusted systems are discussed. Our focus is directed around well-established trusted system hardware approaches (Trusted Computing Group (TCG), specifications) and how they can enhance software tools like virtualization to achieve trust.

The rest of the chapter is structured as follows. In Sect. 6.2, a brief analysis of the security enhancement history through hardware means is discussed. In Sect. 6.3, the notion of trust is defined, various trust establishment models are discussed, and how those models can be used for trusted system establishment is analyzed. In Sect. 6.4, dominant hardware-oriented trusted computing methodologies (based on the trusted platform module (TPM)) are presented. In Sect. 6.5, virtualization technology and its combination with hardware trusted computing are discussed. Section 6.6 concludes the chapter and suggests future directions on trusted system development.

6.2 Hardware Security Modules

IT strong security can be achieved by enhancing existing applications or developing new ones with appropriate built-in secure software or hardware fences. However, the vulnerability of software solutions on malicious manipulation that can bypass software security as well as the slow response of software solutions to security requests have convinced IT security experts that hardware solutions are more appropriate for high security demanding applications like those in financial, military, or governmental environments. This belief led to the development of a wide range of special purpose hardware token devices that act as security arbiters and/or user authentication tools (validating a user's identity by providing a token that only the user possesses). These tokens are based on dedicated hardware processing units consisting of physically tamper-resistant embedded cryptographic processors that communicate with the conventional general purpose system processor in order to offer a predefined set of cryptographic and security services [2].

The first commercial uses of cryptographic processors or hardware security modules (HSM) were made for financial transactions. In such applications HSMs enforced a policy on key usage along with a variety of key protection measures. Electronic payment systems use the HSMs for secure communication between the banks and the customers and for secure storage of all authentication information. The customer is provided with a cheap autonomous HSM (smart card) along with a personal identification number (PIN) for authentication. This smart card solution guarantees end-to-end security in the communication between the bank and its clients.

The introduction of Internet banking brought new dynamics in the field of financial transactions since the customer has no physical presence in a prearranged place to use the bank services. To access bank services ubiquitously through internet banking, the user-customer needs to build and maintain a secure, trusted environment, irrespective of his physical location. Banks currently address this challenge by providing to their customers tamper-resistant authentication, i.e., authorization devices (e.g., the RSA SecurID) that can generate time-dependent or random passwords based on unique registered key in the device. However, the customer is not provided with any safeguards of validating his internet banking access device (laptop, desktop, tablet). Customers must apply their own additional security measures (e.g., firewall, antivirus, antimalware software) in order to trust their access devices while the bank itself always considers these devices untrusted.

HSMs are widely used in military–government applications. Military cryptographic processors have been used from the Cold War era for encrypting sensitive communications and for authorizing people as well as for protecting high-importance military operations. Some of these technologies have been replicated in crisis management situations where civil protection agencies (police, fire brigade, and ambulance staff) need to communicate over secure channels. Proprietary secure communication channels have been used in such scenarios (TETRA, Tetrapol, etc.[3]) that strongly rely on dedicated HSMs in an attempt to create a secure communication environment over an untrusted infrastructure (wireless links, telephone network, etc.).

6.3 Trust

While security might be the dominant term when it comes to protection of sensitive data, trust is a much stronger concept that goes beyond confidentiality, availability, integrity, and nonrepudiation (the basic security pillars). Trust tries to formulate a good-faith relationship between computing machines as well as between their users. The realization of trusting an entity B by an entity A is based on the belief that B will always behave honorably, reliably, and securely under a specific context [4]. From IT perspective, trust is not only about securing the communication channel or authenticating the data sender but also on trusting that the sent information are legitimate, they do not include malicious codes (e.g., malicious, virus, or trojan code) and they will not harm the receiver in an unforeseen way. In other words, trust extends to the sender itself (not just its messages) by believing that he will obey to specific communication rules (dictated by the communication protocol or policy) and will not abuse communication by nonresponsiveness or selfish behavior.

6.3.1 Trust Establishment Models

Establishing trust in computer systems must involve both the user and the computing device at hand. This can be achieved by using a "hard evidence" approach where (1) the device always complies with specific rules thus becoming predictable and (2) the device user behavior conforms to prescribed rules favoring a behavior pattern (or series of patterns) that can be considered legitimate. The series of device rules constitute a specific trust policy which can be also extended to a user behavior policy. To achieve this, appropriate monitoring mechanisms are required so as to validate the computer system (device and user) compliance to the trust policy. The system is considered trusted when the policy is followed and the user behaves in a trusted manner. Under this view, however, it has to be considered that the enforcement of the trust policy to all involved parties necessitates a trust mechanism that is hard to implement especially when users are involved. Thus, alternative approaches must be adopted, which are based on "trust reputation evidence." Reputation monitoring mechanisms are focused on assessing the trust level of an entity (the computer user and device) based on the entity's interaction and behavior history within a given context. More specifically, this history, constituting the entity's reputation, is built from the reports and observations collected by other entities (a single entity acts as evaluator or a group of entities act as trusted third parties).

In a trust model dictated by the above directives (i.e., policy and reputation), the communication channel between a sender and a receiver is always considered secure using strong security mechanisms as well as penetration-resistant means. Both sender and receiver gain value in trusting each other which exceed the performance cost of communication. Value is derived when the involved parties have a need for communication. For example, when the receiver needs to use the communicated information to invest on computer resources or when the sender is obliged to respond to a receiver data request. The level of trust that an involved entity A has to another entity B has a communication cost which is translated into the risk generated when entity B behaves in a malicious, unforeseen way. High trust benefit (i.e., high value) comes when entity A trusts B, meaning that the cost associated with communication with B is smaller than the value of the communication outcome (e.g., the message transmission) [5].

To minimize the risk associated with trust relationships, each involved entity can adopt an isolation mechanism to increase the communication value. Isolation is based on the adoption of trust verification mechanisms during communication.

Instead of blindly trusting the sender, a receiver can follow approaches based on policy and/or reputation in order to protect it from malicious or incompetent senders. The approaches extend not only to message verification but also on noncompliance with a communication protocol or nonresponsiveness of protocol participants. As can be derived from the previous analysis, the trust verification and isolation mechanisms can be categorized in policy-based trust approaches and reputation-based trust approaches.

Policy-based trust establishment is focused on the enforcement of a specific policy capable of guaranteeing trust relationships between communicating participants. Strong isolation and high value for each such participant is maintained by collection, management, and verification of policy-related credentials. Such credentials are meant to be provided and verified by a trusted third party (TTP) authority. Collecting enough credentials by a receiver entity constitutes a proof that the sender can be trusted, thus the receiver no longer remains isolated from this sender. Gathered trust information about an entity can be considered such credentials. Policy-based trust establishment includes trust negotiation protocols with requests for trust credentials, TTP credential verification, and generation of trust assurances. During the execution of a negotiation protocol, an entity usually must provide private content information. Sensitive trust-associated data must be handled appropriately so as to protect the entity computer system's and user's privacy. So it is obvious that trust has a tight interdependence with strong security (although it serves different purposes) since trust negotiation as well as credential handling follows security-related approaches based on cryptographic primitives and well-known security protocols. However, trust policies have a well-specified "language" that is associated with the employed trust credentials. Trust relationship representation can have many forms depending on the trust standard that is adopted. The representation language involves software structures, like Web services [4], but requires a hardware base using HSM (trust anchor), acting as trust policy arbiter on the associated computer [6].

In the *reputation-based trust establishment*, the trust experience of a community of other entities is used in order to make trust decisions about a single entity. In this approach, when a receiver wants to know if he can trust a sender, then he "asks" the opinion of a third party to attest the sender's trust level. Based on the collected replies, the receiver infers if the sender is trusted or not [4, 5]. If the third party is a single trusted authority, then the system is decomposed into a policy-based one. Instead of obtaining trustworthiness-related information from a centralized trusted third party, the reputation-based model collected knowledge of many entities that have prior experience with the evaluated entity. The reliability of the above approach is maintained by a trust reputation-recommendation system. Such system relies on the opinion of a series of recommenders and evaluators of trust that are collecting and analyzing behavior patterns of involved entities. The larger the numbers of recommenders, the more reliable are the trust decisions that can be made. Of course, the recommenders themselves must also be trusted if their opinion is to be of any real value.

Reputation systems usually work in decentralized manner and can be applied to networks that favor data collection and distribution such as the IP networks (the

Internet), p2p networks, and grids [4]. Reputation collection is achieved using software tools (agents) roaming the network, collecting trust reputation data from network entities or other agents. Delivering such information to a requesting entity can lead to an appropriate trust decision. In this notion, reputation can be defined as a measure of trust that each entity maintains and shares with other entities thus forming a "Web Of Trust" (WOT). Transferring trust in a WOT approach is based on trust metrics and trust transitivity rules. Entities that have achieved a level of trust have a WOT link thus forming WOT graphs while entities that have no collected trust reputation data follow trust transitivity rules to form a link. Such rules can be expressed in a simplified matter as "if A entity trusts B entity and B entity trusts C entity then A can trust C." Thus traversing the created WOT trust graphs and collecting reputation data (through the trust metrics system) the various agents of WOT can make trust decisions on a system entities. It must be noted that reputation-based model is a fully software solution (a software service) located at the application stack of a computing system (on top of operating system (OS) kernels and hardware structures).

In general, the reputation-based model is primarily used in network security as a means of providing trust to nodes and network resources (network trust) while the policy-based model can more directly be applied to each computer system individually (computer system trust). The focus of this chapter is on ways of providing computer system trust.

6.3.2 *Trusted Systems*

By defining the notion of trust and formulating models to achieve it, a toolbox to build trusted systems is provided to computer security experts. A system can be considered trusted when its functionality is fully predictable or in other words when it always works the way it was designed to work. Breaking this rule will result into critical security problems, since trusted systems are relied upon serious security actions. Therefore, a trusted system is a system that can be trusted not to fail (if it fails the whole security policy collapses). Note that this should be discriminated from trustworthy systems which are systems that cannot fail (it is impossible to fail).

Since trustworthy systems are very difficult (if not impossible) to design, trust and not trustworthiness, is extended to a wide variety of systems covering even nontraditional security (military, business) applications. However, such systems consist of components that are not considered secure (nonsecure hard disk storage, networks, OSs, legacy components, etc.). Replacing those components with secure ones or redesigning them from scratch in an effort to guarantee trust is a very costly operation and is not usually followed. Affordable trusted system security can be achieved by following the policy-based model where trust metrics and rules are managed for all system's components and trusted applications, services, and resources are strongly isolated from untrusted ones. Similar to isolation from external environment mentioned in the previous subsection, computer system component isolation involves hardware/software codesign in an effort to create a trusted section

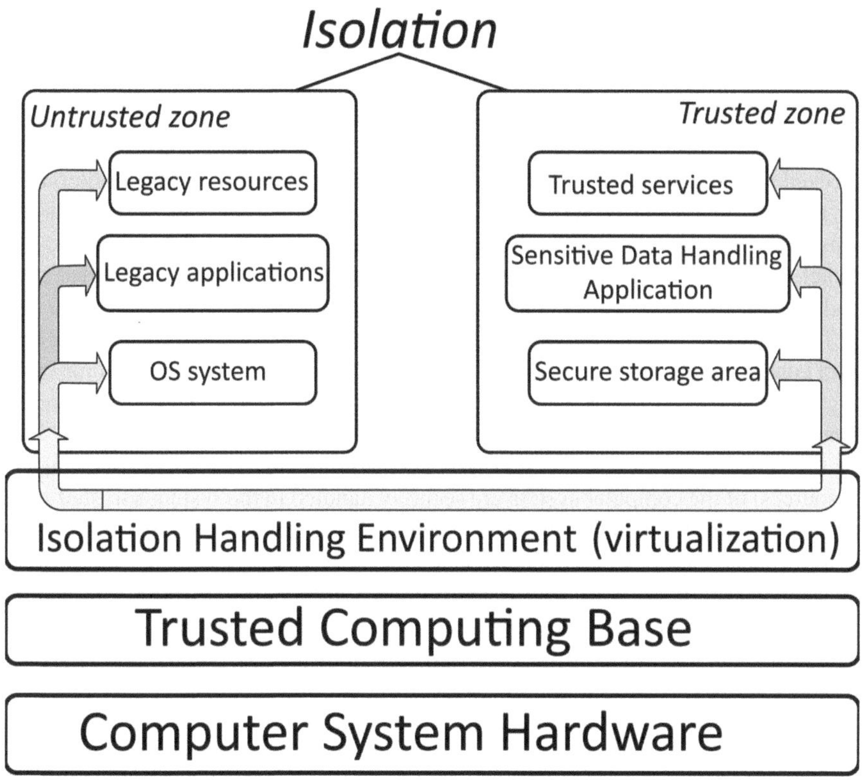

Fig. 6.1 Isolation in trusted systems

within the computer system that is fully reliable for critical applications. Trusted and nontrusted applications are strongly separated between the trusted and untrusted system's zones while their data exchange follows strict security rules (based on the system's policy). This approach can be viewed as an extension of the HSM concept where an HSM is attached externally to a computer system in an effort to provide strong security. In a trusted system, the HSM becomes an integral part of the system's hardware and low-level software, and is assigned the additional role of managing the system's isolation. A common practice on how to achieve this is described in Fig. 6.1.

As described in Fig. 6.1, the system's hardware does not communicate directly with the OS kernel but is rather managed by a trusted computing base (TCB). TCB is a collection of hardware, firmware, and/or software components critical for the reliable functionality of a computer system [7]. As the term suggests, TCB must be trusted since it has the highest OS privilege level and is responsible for the system's security policy enforcement. It includes security validation as well as domain separation mechanisms so as to fully control the information flow of the computer system, provide access control and resource management. The TCB assets and its sensitive data and services must be protected even from itself. To achieve this, the

TCB has special hardware and software protection mechanisms such as protected storage, protected memory, special memory managements units (MMU), TCB software security checkers, and validators in an effort to guarantee trust. On the other hand, for a practical implementation, the TCB must be small enough in order not to burden the system's functionality, to be manageable, and to be efficiently checked periodically from certification organizations, such as Common Criteria [8] for trust (formal security verification). It must be noted, however, that TCBs are considered "trusted" and not fully trustworthy. Security bugs related with the OS kernel (which in some cases is part of the TCB) cannot always be detected thus leading to TCB security risks compromising its trustworthiness.

The isolation between the trusted and untrusted zone is handled by a specialized level consisting of a collection of software tools that use the TCB services for splitting OS resources, services, and applications into trusted and untrusted ones and of controlling/managing all transactions between the two zones. This control is vital for the computer system trust. Trusted subjects (application services) and objects (resources) of the computer system are securely handled in the system's trusted zone and calls or data exchanges with untrusted subjects and objects are not done directly. All such activities must be evaluated and validated by the isolation handling environment acting as a security arbiter between zones. Virtualization is a widely used technology for such isolation handling. Each zone is executed in a different virtual machine environment in order to remain totally autonomous. Alternatively, microkernels [9] are used to achieve the same goal.

A similar solution is offered by ARM in its latest processors (ARMTrustZone technology). ARMTrustZone [10, 11] offers a trust zone environment within the processor structure and not the computer system in general, by splitting processor functionality between a secure and nonsecure operation section, thus making possible the realization of a secure and trusted anchor within its hardware core. So, a designer is able to use the secure part of the processor for implementing trusted services. However, when using the ARMTrustZone or software-based isolation handling environment, it is hard to determine whether such implementations provide fully shielded-secure locations or protected capabilities.

The above approaches are focused on the computer system side and not on the user side. They follow the policy-based trust model but do not involve the user's behavior. This fact can be a security risk for the system as a whole since a malicious insider user can still compromise the system by deactivating trusted services or hacking the software arbitrating the trust enforcement.

6.4 Trusted Computing

In order to cope with the growing need for highly reliable trust enforcement that involves both the computer system and its user, Trusted Computing Platform Alliance (TCPA), an industry alliance was formed in October 1999 in an effort to develop and standardize trusted platform technology. TCPA and its successor, the TCG consortium (established in 2003), are responsible for formalizing, applying, and extending the trusted computing ideas to established computer systems either by introducing new hardware and software modules or by proposing appropriate protocols and directives. Trusted Computing can be viewed as a collection of technologies capable of constantly monitoring the behavior of a given computer system and its user for uncovering a possible compromise and an unexpected behavior. The TCG's view on trusted computing focuses on protecting the system from malicious entities even from its own user. As a result of its work, the TCG consortium has devised a series of specifications for new structures within a computer system that can provide reliable trust establishment. TCG goal is to enforce trust on a system by prohibiting the execution of malicious code, by protecting sensitive data (mainly private keys), and by attesting the system's trust level to other entities. This is achieved by a constant evaluation of the computer system's security from boot time. Since software-based security validation is not fully protected from malicious code injection and hacking attacks, TCG solution is based on hardware protection mechanisms along with software tools to establish trust. Embodiment of this approach is the specification of an HSM structure denoted as Trusted Platform Module (TPM) that is capable of acting as trust anchor within a computer system [1, 12].

6.4.1 Trusted Platform Module

The TPM is a smart-card like HSM chip bound to the computer system (usually soldered on the system motherboard). It acts as a hardware trust anchor (extending the trust zone concept) in order to enforce a trust policy by providing secure storage, public key authentication functionality, integrity measurements for all computer system resources/services as well as trust attestation. TPM functionality plays a key role in at least 16 special purpose platform configuration registers (PCRs) designed to store trust configuration measurements associated with specific computer system resources. PCRs cannot be directly written, they can only store an extended version of their previous value (a hashing of their previous value). Collecting a new trust-related measurement about a specific computer resource (e.g., service, private key) associated with a specific PCR leads to an update of the PCR value (extend operation) based on Eq. 6.1, where $H^{(SHA-1)}()$ is the SHA-1 Hash function and $(+)$ refers to concatenation.

$$PCR \leftarrow H^{(SHA\text{-}1)}(PCR + trust\ \ measurement) \tag{6.1}$$

The extend operation inherits the benefits of hash functions. There are no two identical PCR values coming from different measurements. The operation also provides indication of the order in which measurements were taken and hashed (a hash chain approach). The number of measurements is unlimited since the outcome will always be of the same bit-length. Currently, the TPM 1.2 version uses SHA-1 function and the PCR length is 160 bits.

The TPM is characterized and identified by a set of asymmetric cryptography keys denoted as endorsement key (EK) along with a certificate from the manufacturer.

This set of keys are securely stored in the TPM during manufacturing, they are unique for each TPM chip and are very restrictively accessed (private key cannot be released outside the TPM chip, it is nonmigratable). The EK is associated with three credentials provided usually by the TPM manufacturer and a third-party testing laboratory attesting that the current TPM conforms to TCG specifications, that the TPM is genuine and that the host system is an instantiation of a TPM equipped platform. The EK is used when a user wants to prove TPM generated keys have been produced by a genuine TPM. Therefore, the EK is utilized for certificate decryption of other TPM-generated keys, and this can only take place upon the request of the owner of the TPM cooperating with a certificate authority (CA).

The TPM generates a variety of different asymmetric and symmetric cryptography keys in order to realize its various functions. Such keys are Attestation Integrity Keys (AIK), the storage root key (SRK), and consecutive storage keys stemming from the SRK. All such keys can be either migratable or nonmigratable. Migratable keys can be moved to other TPM chips if there is some problem with a host TPM. Nonmigratable keys cannot be moved to another TPM chip. In TPM 1.2 version, all asymmetric cryptography keys are 2048 bit RSA keys.

Communication with the TPM is typically handled by the TCG device driver library (TDDL) whose interface is described in the TCG software stack (TSS) specification [13]. This software library installed on the TPM host computer system communicates with a device driver inside the TPM kernel. The device driver is responsible for the actual transaction with the TPM hardware device and its functions. The TSS library has all the tools (services, commands) for an application architect to use the TPM for adding strong security features to a software application. TSS functionality can be divided into three logical components: the TDDL, the TCG core service (TCS), and the TCG service provider (TSP). The TDDL provides an API for interfacing the TPM functions. The TCS, being the main user of the TDDL, manages the TPM resources, converts API request to TPM byte streams in order to be recognized by the hardware, and provides system-level key storage (outsize the TPM) while synchronizing application-specific calls coming from the TSP. The TSP is the interface with which applications communicate to the TPM. It offers access to the TPM services transparently for the application, acting as a shared object or a dynamic linked library (dll).

In its current version (TPM 1.2), the TPM chip is equipped with all the necessary hardware components in order to support strong security features. Apart from the I/O interface, necessary for the TPM communication with the external world, inside the TPM there are a series of cryptographic hardware components including a true random generator unit, an asymmetric key cryptography digital signature and authentication-authorization unit, and a hash function unit. Currently, the TPM adopts the RSA algorithm with 2048 bit keys and 160 bit hashing through SHA-1 and HMAC algorithm. The TPM also supports the secure storage of sensitive values (such as asymmetric and symmetric cryptography keys or measurement states) in special storage elements. Those elements include nonvolatile and secure volatile memory as well as a series of at least 16 PCRs [12].

6.4.2 *Trusted Computing Functions and Services*

TCG has specified a number of innovative ideas capable of setting up a fully controllable and trusted computer system environment for the system's user. TCG also describes a mechanism of providing evidence of trust to third parties as well as authentication/authorization. In this section, the basic TPM-TSS functions for achieving the above concepts are described.

6.4.2.1 Authenticated—Secure Boot

Through the TCG TPM mechanisms, the trust state of a system can be reliably measured and recorded. To achieve that, every part of the computer system from hardware level to application level is measured from boot time. This authenticated boot sequence measurement provides the guarantee that the system is not compromised and can be trusted. The TPM PCRs play an important role in the above boot sequence. The outcome of each system's component measurement is stored in a PCR register, according to Eq. 6.1. A recording of this process is stored in a history file denoted as Stored Measurement Log (SML), which is maintained outside the TPM structure.

The authenticated boot measurement follows a daisy chain approach, meaning that each component is measured and compared with existing known good value on the SML. In that way, measurement integrity is retained and the boot sequence can be trusted, meaning that no system component has been tampered. For this approach to work correctly, the boot sequence control must be given to a trusted source. Thus at computer system power-on, control is given to a TPM small root of trust module (a subset of the BIOS) before loading the BIOS. The root of trust takes the first measurement (BIOS measurement), compresses it (creating a digest), and compares it with the stored value inside the history file (SML). If the two values match, then the BIOS component is considered trusted and is executed, a specific PCR (PCR_0) value is extended with the collected measurement and the next component is evaluated in a similar way. The history file's integrity can be guaranteed by comparing it with the PCR values (that cannot be tampered). An overview of the above procedure, denoted as TPM static root of trust, is presented in Fig. 6.2. Note, that every component of the computer system is measured and evaluated in the above fashion including all executable code (i.e., system applications).

Static root of trust has some drawbacks associated with the need for chain trust measurement of every computer system software structure (inclusivity problem). The size of the executable code to be checked can be overwhelming and in a complex system can render its security properties unverifiable (no scalability). Executing a software code using a configuration file or processing data that has not been already measured can break the trust chain and harm the system's trust especially if such code is tampered at the time interval between trust measurements (scalability and measurement time problem). For this reason, static root of trust is used in contained, well-structured computer systems with well-defined operations. Instead, in complex systems requiring hundreds or thousands of measurements in a static root of trust chain, the dynamic root of trust chain approach is used which can dramatically

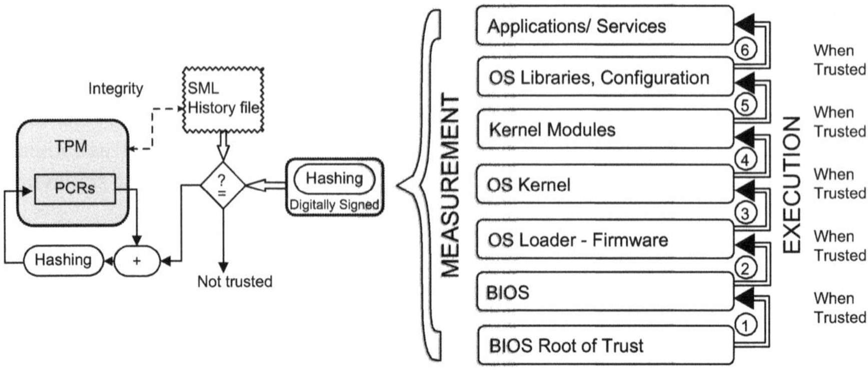

Fig. 6.2 Trusted platform module (TPM) static chain of trust overview

reduced measurement number. Dynamic root of trust (DRT) uses a sophisticated measurement mechanism in order to evaluate the computer system trust level that can be initiated at any point in time and can be repeated as often as necessary. DRT measurement is initiated by a specialized processor instruction creating a secure, attested execution environment [1, 14]. This environment is completely isolated from the rest of the system (e.g., Direct Memory Access is inhibited, interrupts, and virtual memory are disabled) and guarantees untampered execution of a secure loader that initiates the trust measurement process of any executed code. Each processor manufacturer uses a different approach on how to implement the DRT measurement instructions [15, 16].

6.4.2.2 Secure Storage

The TPM chip is capable of storing a wide variety of information in a secure way. Such information can be divided into asymmetric cryptography keys and symmetric cryptography keys or data. Secure storage follows a protected object hierarchy where higher-level keys are used for signing-protecting lower-level keys within this hierarchy. The root of the key hierarchy is the SRK which is an asymmetric key pair (2048 bit keys RSA keys) generated using the EK at the first TPM power-on. The private key of SRK must never leave the TPM and is therefore nonmigratable. Then, the SRK can be used along with the TPM random number generator in order to create storage keys for each piece of information needed to be securely stored. The storage keys are located into the key hierarchy and are protected by higher level keys. In addition, the TPM key generation mechanism can provide keys that are used only by the TPM that generated them (applies to private keys) and/or when the TPM's host platform is in a specified state. Securely storing keys or data is done using asymmetric key cryptography for confidentiality-integrity and can be associated with a 160-bit string of data in order to provide authentication before encryption.

The TPM can be used in order to provide secure sealing functionality. The encrypted information can be sealed so that they can only be decrypted by the TPM used for their encryption and only when the TPM Host platform is in a specified state. Practically, this can be done by performing asymmetric cryptography encryption/decryption using the SRK (that is unique for each TPM chip) and associating the encryption/decryption with specified values of some PCR.

6.4.2.3 Platform Remote Attestation

A very important TPM feature is the ability to provide TPM host system trust attestation reports to external third parties thus proving that the system can be trusted. The attestation operation should be unique for each TPM host computer system and undeniable. The TPM can be uniquely identified using the EK values. However, the need of protecting the TPM identity and its host's anonymity dictates that this approach cannot be used in practice. When the TPM host, denoted as Trusted Platform (TP), must provide attestation of its trust level to a third party, its anonymity must be retained so that its activities cannot be tracked. For this reason, the TPM generates a series of pseudonyms in the form of asymmetric cryptography key pairs, denoted as attestation identity keys (AIK), in association with the host system and a Certificate Authority (privacy certificate authority (PCA)) that provides AIK credentials. To achieve that, the TP provides the EK credentials and AIKs to the PCA, the PCA verifies that these credentials are legitimate thus the TP is genuine, and then generates an AIK credential by digitally signing (binding) the AIK public key with the description of the TP [1, 14]. In this process, the identity of the TP and the TPM (i.e., the EK) should not be revealed by the PCA. For this reason, the TCG has specified the Direct Anonymous Attestation protocol using provable security features and zero knowledge protocol cryptographic approaches to retain privacy [13, 17].

The attestation process, denoted as remote attestation, involves a specific AIK key pair, a TPM-specific state (denoted TPM quote) that provides a captured instant of the PCRs values and a series of nonce numbers. When an entity wants to have insurances about the trust level of the TP, it sends a request to the TP along with a nonce. In return the TPM sends back a digitally signed by the AIK private key, concatenation of the nonce and the TPM quote along with the appropriate AIK credential and a section of the SML. The requesting entity verifies the AIK credentials and digital signature as well as the nonce value that it originally sent to the TP and acquires the TPM quote. It then compares the TPM quote and the metrics provided from the SML with trusted known good values stored in trusted third-party database and if the values match then the TP is considered trusted.

6.4.2.4 Trusted System Realization with Trusted Computing Group Specifications

The trusted system concept, as described in Sect. 6.3.2, applies to the TCG specifications about trust. The hardware TPM structure provides an untampered, secure environment for supporting a TCB. Such TCB can be measured for trust and remain trusted at all times, thus, providing the basis for isolation. The TPM and TSS not only provide the means for evaluating trust for a computer system but also indirectly for the system user. They can guard the TP even from its own user since they do not allow malicious behavior (code injection, illegitimate software modifications) through the "secure boot" feature. The TCG approach can effectively enforce a TCG specified policy on the computer system and its user by hardware means and through measurement provide, in a way, a "reputation" collection system stored in the PCRs and SML. This system collects trust "reputation" locally about the various system components. This local reputation can be transmitted to other entities through remote attestation. This function requires that a TPM TP's "reputation" data (TP trust measurements) is being stored in trusted third parties within a computer network (e.g., using Internet) thus structuring and retaining "reputation" databases. This approach can be viewed as a form of static reputation collection mechanism following the reputation-based trust model. The TCG has specified ways of enhancing this direction through the TCG network connect (TNC) scheme which is beyond the scope of this chapter. Furthermore, several researchers have proposed ways of using TPMs for securing mobile agents roaming a network for data collection that also can be used for reputation collection [18–22].

TCG tries to provide a trust establishment solution that uses concepts from both the policy-based trust model and the reputation-based model, infusing them to an HSM (the TPM) characterizing the computer system (since it is soldiered to it) as well as creating all the HSM necessary supported software.

The TCG solution has guided the implementation by top hardware vendors (currently Intel and AMD only) of an isolated execution environment (IEE) (through processor instruction) to be used originally for DRT measurement. This environment offers isolation at hardware level and is inline with Fig. 6.1 concept. Setting up one independent IEE for each application or group of applications can provide strong isolation since the TPM and TSS can guarantee trusted communication between environments and the external world.

6.5 Virtualization Environment for Trust

Virtualization is a technology with high potentials in securing the computing world and providing trust. It provides an abstraction of one computer system level to another higher level. Virtualization can be found in many forms from network systems to storage or process virtualization. However, from security perspective, it is especially interesting to view these technology actors as reference monitor mediators of access to system resources and communication between abstraction environments within the system. These types of actors are referred as virtual machine monitors (VMM) or hypervisors while the abstraction environments are denoted as virtual machines (VM). A Hypervisor can be a small software code, positioned between the hardware and the OS kernel computer system levels in a similar way as the Isolation Handling Environment presented in Fig. 6.1. By introducing a VM environment where not only

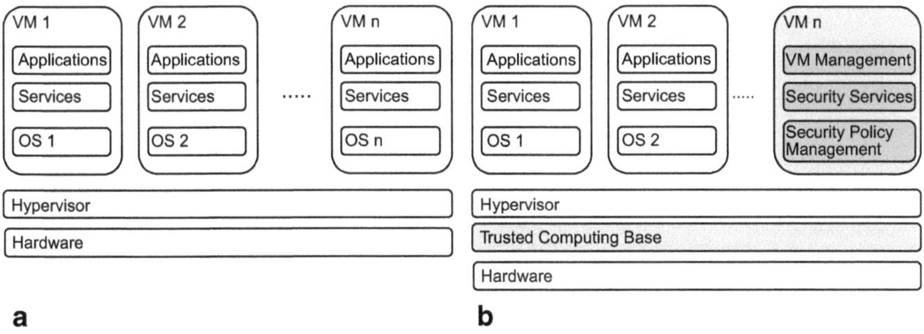

Fig. 6.3 Computer system architecture featuring hypervisors

programs but also OSs can run as if executed on hardware, hypervisors can achieve strong isolation. Thus, logical or network attacks on an OS application executed on top of a VM can be contained there and won't spread to the rest of the system. Further security advantages of hypervisor systems can be the ability to cryptographically protect data under a contained environment (cryptographic separation), the running of workloads at different time periods (temporal separation), and the assignment of specified hardware resources to different VMs (physical separation) [23]. Figure 6.3 describes generic approaches on security-oriented hypervisor-based systems.

Hypervisors can be used in cooperation with the TPM chip in order to take advantage of TCG trust functionality. Virtualization gives further dynamic to TPM trusted systems since it sponsors strong isolation beyond the TPM. On the other hand, TPM can be used to provide a secure hardware virtualization interface, practically forming a virtualization enabled TCB. However, the TPM cannot directly be used from VMs since it was designed for a single host platform. Only one such platform can have access to the TPM hardware structures and especially the PCRs. In a virtualization scenario, each VM can act as an independent platform and can potentially change the PCR of the single TPM attached to the host system. In that case, there is a serious security danger since VM trust states assigned to a PCR can be changed by a different VM. Several researchers have suggested solutions to this problem by adopting the concept of a software virtual TPM (vTPM) instance residing in each VM (as depicted in Fig. 6.4a) and communicating with the TPM structure in a manageable way through the hypervisor. Terra [24] was one of the first systems to introduce trusted computing to hypervisors (although without the use of TPMs) and has been followed by several other works [25–27] that employ the vTPM concept. Another approach introduced in sHype hypervisor [28] is presented in Fig. 6.4b. In this approach the vTPM is running on a dedicated VM and communicates through the hypervisor with a hardware TPM. The rest of the VMs have only vTPM drivers in order to communicate with the hypervisor trust interface. sHype also supports a dedicated VM for the system's security policy management and an access control mechanism inside the hypervisor [23].

Other approaches related to virtualization include the employment of microkernels such as the L4 [29] or seL4 [30, 31] or microvisors such as OKL4 [32] instead of

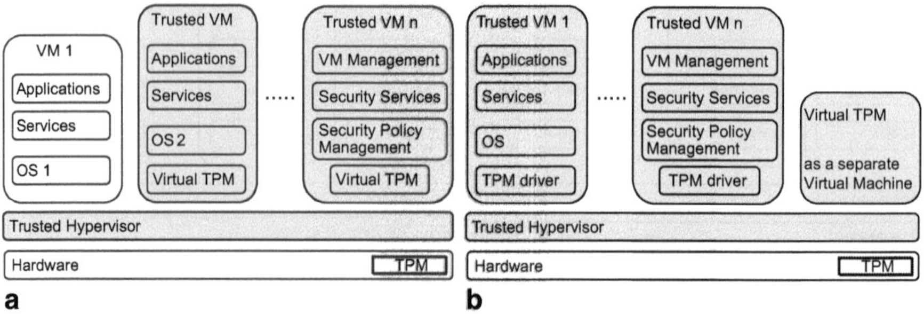

Fig. 6.4 Trusted platform module (TPM)-based hypervisor systems

hypervisors for achieving isolation. In such cases, the microkernel/microvisor must be an extension of the TCB in order to be trusted and executes security critical operations.

Virtualization has also been introduced in processor architecture level, as briefly mentioned in the previous section (DRT measurement), by extending the processor instruction set with special virtualization instructions in Intel or AMD processors through the trusted execution technology (TXT) [15, 33] and secure virtual machine (SVM) [16, 34] technology, respectively. The two approaches are implemented in a different way but both provide the means of creating VMs through totally hardware measures. The collaboration of the TXT or SVM technology with the TCG TPM enables DRT measurement and provides IEE as described in previous sections.

Using any of the above virtualization approaches, a system security designer can create trusted areas of virtual machines running on virtualizated hardware and direct sensitive applications and data toward those virtual machines. Extending this logic, trusted OS can be run on such VM and as long as access is controlled by a TCB program on the processor, the OS remains isolated and protected from the rest of the system's untrusted VMs. Trusted virtualization technology is still on an early stage, since several practical implementation problems still exist (hardware constrains, system real-time behavior, scheduling, access control rights). However, virtualization opens the road for unlimited hardware/software codesigned structures that through HSMs like the TCG TPM, can lead to very high trust-level systems.

6.6 Conclusions and Future Directions

In this chapter, the road toward designing hardware-based trusted system was described, beginning from monolithic implementations of HSM for specific applications and moving toward very dynamic solutions like the TPM offering sophisticated functionality to associated software. Parallel to this approach, virtualization was evolved over time into a very useful security enhancement tool. Thus, the merging of the hardware security world with software virtualization has resulted in very strong isolation mechanisms capable of providing high trust level. In the future, this approach is bound to be expanded and further adopted. TPM chips will migrate from the desktop, notebook, and server computer domain to the mobile world enabling us to, ultimately, trust our mobile devices (smart phones, tablets) for security-sensitive transactions. The TCG has moved toward this direction by providing specifications for a mobile TPM version (Mobile Trusted Module, MTM) but these specs have not led to a market product yet. Furthermore, the growing adoption of embedded computing systems in everyday devices has stemmed the need for strong security and trust. Trusted computing will play a very dynamic role in securing the embedded system world and TPM structures along with virtualization will, eventually, be infused into embedded systems.

On the other hand, the current TCG specifications (TPM v1.2) on the TPM have several shortcomings since they lack flexibility and diversity. Already, TCG is working on changing that by providing a TPM v2.0 (a preliminary draft was released in October 2012). From cryptographic perspective, designers need to put behind the aging cryptographic infrastructure of the existing TPM and adopt more efficient cryptographic schemes. Researchers point to Elliptic Curve cryptography (ECC) as the most suitable candidate for a public key scheme and to SHA-256 or the upcoming SHA-3 as the most suitable Hash function scheme [35, 36]. The presence of an ECC framework can provide additional TPM features such as ECC pairing-based cryptography (PBC) (using Weil pairing, Tate pairing, Eta pairing, Ate pairing, etc.) capable of supporting advanced security services such as short signatures, identity-based encryption and signature, identity-based authenticated key agreement, Tripartite Diffie-Hellman or self-blindable credentials [37]. The Direct Anonymous Attestation (DAA) mechanism, adopted by TCG, that is currently based on symmetric pairings can be more efficiently realized using ECC asymmetric pairings [38].

Finally, the wide adoption of multicore processors will have a profound impact on trusted computing and virtualization. Apart from high-speed implementations, dynamic use of multiple cores will enable better hardware protection patterns. Hypervisors will be able to assign whole VMs to specified cores and provide hardware virtualization on processor core level.

References

1. Challener, D., Yoder, K., Catherman, R., Safford, D., & Van Doorn, L. (2007). *A practical guide to trusted computing*. Boston: IBM press.
2. Anderson, R., Bond, M., Clulow, J., & Skorobogatov, S. (2006). Cryptographic processors-a survey. *Proceedings of the IEEE, vol. 94*, 2, pp. 357–369.
3. PracTel Inc. (2007). Tetra and tetrapol: Technology and market comparison. PracTel Inc.
4. Artz, D., & Gil, Y. (2007). A survey of trust in computer science and the semantic web. *Web Semantics: Science, Services and Agents on the World Wide Web, 5*(2), 58–71.
5. Gligor, V., & Wing, J. M. (2011). Towards a theory of trust in networks of humans and computers. *Security Protocols XIX. Springer*, pp. 223–242.
6. Aussel, R. B., & Sailer, J. D. A. (2011). Only hardware-assisted protection can deliver durable secure foundations. *IEEE Software, 28*, 2, pp. 57–59.

7. Rushby, J. (1984). A trusted computing base for embedded systems. *Proceedings 7th DoD/NBS Computer Security Initiative Conference, Gaithersburg, MD*, Sep. 1984, pp. 294–311.
8. Criteria, C. Online. http://www.commoncriteriaportal.org.
9. Iqbal, A., Sadeque, N., & Mutia, R. I. (2009). An overview of microkernel, hypervisor and microvisor virtualization approaches for embedded systems. *Report, Department of Electrical and Information Technology, Lund University, Sweden, 2110.*
10. Alves, T., & Felton, D. (2004). Trustzone: Integrated hardware and software security (ARM white paper). *Information Quarterly, 3*(4), 18–24.
11. Armtrustzone, A. R. M. http://www.arm.com/products/processors/technologies/trustzone.php.
12. Group, T. C. (2007). TCG TPM specification version 1.2. https:/www.trustedcomputinggroup.org/specs/TPM/.
13. Group, T. C. (2006). TCG software stack (tss) specification version 1.2. http://www.trustedcomputinggroup.org/resources.
14. Fisher, D. A., McCune, J. M., & Andrews, A. D. (2011). Trust and trusted computing platforms. DTIC Document, Tech. Rep., 2011.
15. Intel. (2011). Intel trusted execution technology (intel txt).
16. Devices, A. M. (2005). AMD, secure virtual machine architecture reference manual.
17. Chen, L., Morrissey, P., & Smart, N. (2008). Pairings in trusted computing. *Proceeding of Pairing-Based Cryptography Pairing*, pp. 1–17.
18. Shen, Z., & Wu, X. (2008). A Trusted Computing Technology Enabled Mobile Agent System. *Computer Science and Software Engineering, International Conference on, 3*, pp. 567–570.
19. Wilhelm, U., Staamann, S., & Buttyan, L. (1998). On the Problem of Trust in Mobile Agent Systems. *Internet Society's Symposium on Network and Distributed System Security*.
20. Tan, H. K., & Moreau, L. (2001). Trust Relationships in a Mobile Agent System. *Mobile Agents, number 2240 in LNCS, Springer*, pp. 15–30.
21. Uwe, S. S., Wilhelm, G., & Buttyan, L. (1999). Introducing Trusted Third Parties to the Mobile Agent Paradigm. *Secure Internet Programming: Security Issues for Mobile and Distributed Objects, Springer-Verlag*, pp. 471–491.
22. Hein, D., & Toegl, R. (2009). An autonomous attestation token to secure mobile agents in disaster response. *The First International ICST Conference on Security and Privacy in Mobile Information and Communication Systems (MobiSec 2009)*. Torino, 2009. From HST to Trusted Computing and Trusted systems 19.
23. Perez, R., van Doorn, L., & Sailer, R. (2008). Virtualization and hardware-based security. *Security & Privacy, IEEE, 6, 5*, pp. 24–31.
24. Garfinkel, T., Pfaff, B., Chow, J., Rosenblum, M., & Boneh, D. (2003). Terra: A virtual machine-based platform for trusted computing. *ACM SIGOPS Operating Systems Review, 37, 5. ACM*, pp. 193–206.
25. Berger, S., Caceres, R., Goldman, K., Perez, R., Sailer, R., & van Doorn, L. (2006). vTPM: Virtualizing the trusted platform module. *Proceedings of 15th Conf. on USENIX Security Symposium*, pp. 305–320.
26. Stumpf, F., Benz, M., Hermanowski, M., & Eckert, C. (2007). An approach to a trustworthy system architecture using virtualization. *Autonomic and Trusted Computing, Springer*, pp. 191–202.
27. Stumpf, F., & Eckert, C. (2008). Enhancing trusted platform modules with hardware-based virtualization techniques. *Emerging Security Information, Systems and Technologies, 2008. SECURWARE' 08. Second International Conference on IEEE*, pp. 1–9.
28. Sailer, R., Valdez, E., Jaeger, T., Perez, R., Van Doorn, L., Griffin, J. L., & Berger, S. (2005). sHype: Secure hypervisor approach to trusted virtualized systems. *Techn. Rep. RC23511.*
29. Härtig, H., Hohmuth, M., Liedtke, J., Wolter, J., & Schönberg, S. (1997). The performance of μ-kernel-based systems. *ACM SIGOPS Operating Systems Review, 31, 5, ACM*, pp. 66–77.
30. Heiser, G. (2008). The role of virtualization in embedded systems. *Proceedings of the 1st workshop on Isolation and integration in embedded systems, ACM*, pp. 11–16.
31. Heiser, G., Andronick, J., Elphinstone, K., Klein, G., Kuz, I., & Ryzhyk, L. (2010). The road to trustworthy systems. *Proceedings of the fifth ACM workshop on Scalable trusted computing, ACM*, pp. 3–10.

32. Heiser, G., & Leslie, B. (2010). The okl4 microvisor: Convergence point of microkernels and hypervisors. *Proceedings of the first ACM asia-pacific workshop on Workshop on systems, ACM*, pp. 19–24.
33. Uhlig, R., Neiger, G., Rodgers, D., Santoni, A. L., Martins, F. C., Anderson, A. V., Bennett, S. M., Kagi, A., Leung, F. H., & Smith, L. (2005). Intel virtualization technology. *Computer, 38, 5*, pp. 48–56.
34. Strongin, G. (2005). Trusted computing using amd pacifica and presidio secure virtual machine technology. *Information Security Tech. Report, 10, 2*, pp. 120–132.
35. Zhang, X., Zhou, M., Zhuang, J., & Li, J. (2007). Implementation of ECC-Based Trusted Platform Module. *Machine Learning and Cybernetics, 2007 International Conference on, 4, August, IEEE*, pp. 2168–2173.
36. Fournaris, A. (2012). Toward flexible security and trust hardware structures for mobile-portable systems. *Latin America Transactions, IEEE (Revista IEEE America Latina), 10, 3*, pp. 1719–1722.
37. Barreto, P. S., Kim, H. Y., Lynn, B., & Scott, M. (2002). Efficient algorithms for pairing-based cryptosystems. *Advances in cryptologyCRYPTO 2002. Springer*, pp. 354–369.
38. Brickell, E., Chen, L., & Li, J. (2008). A new direct anonymous attestation scheme from bilinear maps. *Trusted Computing-Challenges and Applications, Springer*, pp. 166–178.

Chapter 7
Using Codebender and Arduino in Science and Education

V. Georgitzikis and D. Amaxilatis

Abstract Technology and computers are becoming cheaper and easier to find every day, but it still remains hard to learn and understand how things actually work. Arduino is a great example of how a simple, inexpensive, and easy-to-program device can help students of all ages learn electronics and programming in just a few steps. Although such embedded devices and electronics have been adopted by the community, the barrier of entry remains high in comparison to other technologies like web design and generic computer programming. Additionally, collaboration and exchange of ideas remains hard and bound to the past. Internet and cloud technologies provide a solution to the above. Spreading the knowledge and sharpening the learning curve is the target of Codebender. Codebender is an online learning and collaboration hub for makers, students, and engineers. Students can benefit by learning easier and getting to the point where they can actually program much faster than before. Engineers and scientists get access to advanced development tools that help them code and collaborate with their colleagues faster and without pain.

Technology becomes a more important part of our life on a daily basis. Computers, smartphones, and smart appliances get cheaper and easier to use. Also, more computers and electronic devices are installed in our homes, workplaces, cars, or even gardens making our lives simpler and safer. These breakthroughs and the penetration of technology over the past years can easily be compared to the growth of the Internet in the past decades. More and more people start interacting with technology in their environment as people did in the past with the web via their personal web pages, blogs, or social networks. Though in comparison to the Internet, technology is becoming part of our live faster than ever, people are still slowly getting familiar to how such technological devices work and function in the greater scheme of things.

V. Georgitzikis (✉) · D. Amaxilatis
Computer Technology Institute and Press "Diofantus", Patras, Greece
e-mail: georgitzik@ceid.upatras.gr

D. Amaxilatis
e-mail: amaxilat@cti.gr

N. Sklavos et al. (eds.), *System-Level Design Methodologies for Telecommunication*,
DOI 10.1007/978-3-319-00663-5_7,

Fig. 7.1 An Arduino Due board

7.1 Electronics and Arduino

Arduino is a small factor open-source electronics prototyping platform based on flexible, easy-to-use hardware and software. It is intended for artists, designers, hobbyists, students, and anyone interested in creating interactive objects and environments. The hardware consists of an 8-bit Atmel AVR microcontroller but there is also a newer version (Arduino Due) designed around a 32-bit Atmel ARM processor Fig. 7.1.

Most Arduino boards are programmed via USB, using a USB-to-serial adapter provided either directly on chip or using an external circuit. The Arduino board exposes most of the microcontroller's I/O pins for use by other circuits. The Diecimila, Duemilanove, and current Uno provide 14 digital I/O pins, six of which can produce pulse-width modulated signals, and six analog inputs. These pins are on the top of the board, via female 0.1 in. headers. The software is programmed using a very simple programming language based on processing and very similar to basic C++ called "Wiring". A typical Arduino program requires the user to define only two functions to make a runnable cyclic executive program:

- setup(): a function run once at the start of a program that can initialize settings.
- loop0(): a function called repeatedly until the board powers off.

A typical first program for a microcontroller simply blinks an LED on and off. In the Arduino environment, the user might write a program like this:

```
#define LED_PIN 13


void setup () {
  // enable pin 13 for digital output
  pinMode (LED_PIN, OUTPUT);
}


void loop () {
  // turn on the LED
  digitalWrite (LED_PIN, HIGH);
  // wait one second (1000 milliseconds)
  delay (1000);
  // turn off the LED
  digitalWrite (LED_PIN, LOW);
  // wait one second
  delay (1000);
}
```

The original Arduino schematic (details on the circuitry and connections) is available to everyone to study, reuse, modify, or reproduce it. This open source nature helped created a huge community that created a number of compatible devices to address any possible issue or problem based on the original hardware design. On top of the Arduino clones, the family of the official Arduino devices also includes a number of variants like the Arduino Ethernet for projects that require networking, the Pro Mini that can be easily connected to custom printed circuit board (PCBs) and the Leonardo that is capable of being used as keyboard, mouse, or gamepad. Numerous Arduino clones currently exist, like the TinyDuino, Digispark for even smaller in size projects, the WifiDuino for projects that require WiFi connectivity, and the LilyPad, a board designed to be sewed on clothes to create wearable projects (Figs. 7.2, 7.3 and 7.4).

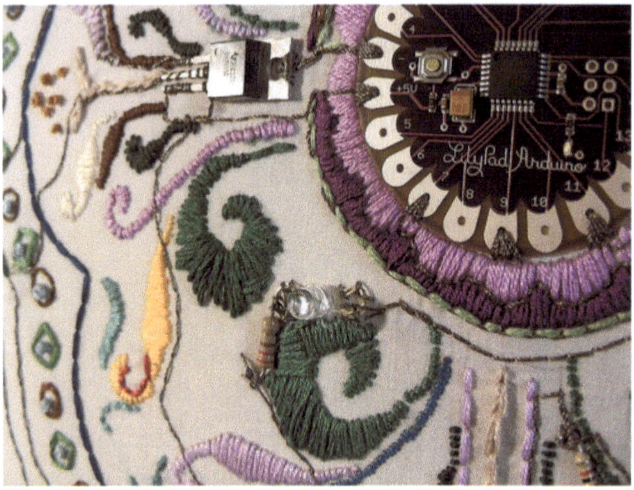

Fig. 7.2 An Arduino LilyPad

Fig. 7.3 Digispark

Fig. 7.4 TinyDuino

Its small size and simple interface makes it appropriate for beginners and amateurs and people who do not have specific knowledge on electronics or software engineering. Arduino offers easy access and connectors to work with and extend them via a huge number of sensors and extensions available off-the-shelf (Figs. 7.5 and 7.6).

Another important thing that led to the wide acceptance of the Arduino from the Open Source and Make communities is the ease of extending the original platform. This has led to numerous extension boards called shields that make it even simpler for someone to add further capabilities to a project. Amongst others, there are shields for

Fig. 7.5 Arduino Uno with an Ethernet shield

Fig. 7.6 Arduino stackable shields

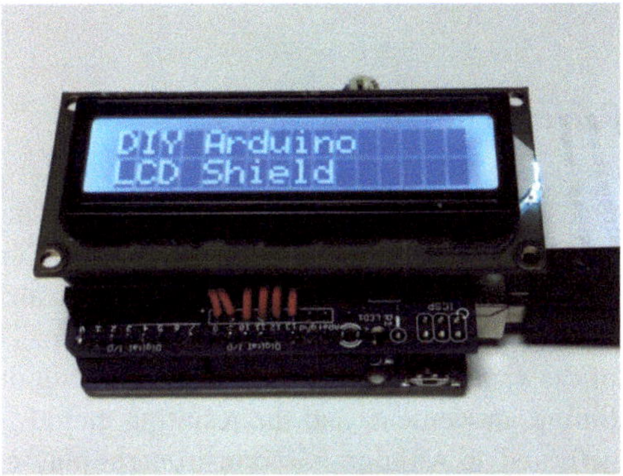

Fig. 7.7 DIY LCD shield

LCD (Fig. 7.7), touch or even electronic ink screens, wireless and wired communication (including ZigBee, XBee, WiFi, and plain Radio Frequencies), relay controllers to control high-power electrical equipment, servo motor controllers to build robots, remote controlled vehicles, or generic actuators. All the available shields are usually accompanied by a code library and different examples that showcase its operation. On top of the examples, videos and connection diagrams and tutorials are widely provided by various blogs and websites all over the Internet (Fig. 7.8).

Fig. 7.8 Arduino stackable shields

7.2 Interactive Installations using Arduino

Apart from being a prototyping device, Arduino is extremely popular to artists designing interactive installations and exhibitions. Its ability to be easily interfaced to a variety of sensors (PIR movement sensors, luminosity and proximity sensors or microphones), actuators like speakers, light bulbs, or small motors, and other integrated circuits (i.e., GPRS, GPS, Ethernet, Bluetooth, and WiFi chips) is what led to this wide range adoption. Many such installations have been exhibited around the world in museums, artistic expositions, and international conferences.

7.2.1 The Tuned Stairs

One of the most characteristic examples of such an installation is the "Tuned Stairs" presented during the Fabrica exhibition at Center Pombidu between 6 October and 13 November 2006 (Fig. 7.9). As visitors walk down the stairs leading to the exhibition, their footsteps activate a musical note. The installation refocuses the attention of the visitor onto their footfalls—allowing them the opportunity to compose their timing, movement, and the resulting melody. On each of the stairs, a matt-switch activated an Arduino-based instrument-playing mechanism attached on the outside of the banister. The armature is made of steel with a pull-thrust solenoid attached to it. Above the solenoid is a tuned resonating bar facing down. As the matt-switch is activated, the solenoid momentarily strikes the resonating bar—resulting in a soft delicate sound resonating through the local area. The matt-switch's signal gets fed into an Arduino board which then activates the solenoid mechanism through a transistor circuit. Each Arduino had its own shield board built to easily connect the switches to their corresponding solenoid mechanisms.

Fig. 7.9 The Tuned Stairs installation

Fig. 7.10 The Garduino installed

7.2.2 Social Gardening

Another example is the Garduino project (Figs. 7.10 and 7.11). Garduino is a complete solution for serving the needs of your plants. What it offers is an open source solution that allows you to water your plants when they are thirsty, turns on supplemental lights when the sun is not out for long, and sends you alerts when temperatures are botanically uncomfortably chilly so you can avoid any unpleasant accidents. As an open source project, all designs and codes are available for anyone interested in reproducing it.

The original project contained a number of sensors including temperature, soil humidity and luminosity, controllers for the water supply and lamps, as well as different methods for sending information to the owner either via the Internet

Fig. 7.11 The Garduino kit

(like email or even twitter) or SMS messages. Many similar projects have been presented over the last years such as the tweeting plant, the Arduino based aquaponics, and many others.

7.3 Communities and Open Source

First and foremost, the Arduino has its own, vibrant online community. It is estimated that there are millions of Arduino users around the world from hobbyists, who use it to design DIY devices to professional developers alike, who use it to increase their productivity and effectiveness when designing commercial electronics devices.

Due to its open source nature, Arduino has become a favorite tool among open source (hardware and software) and various online communities. For example, it is the preferred tool of choice among the Maker movement, which is a subculture representing a technology-based extension of DIY movement. Typical interests include electronics, robotics, 3D printing as well as more traditional activities such as metalworking, woodworking, and traditional arts. It is therefore easy to understand why the Maker movement has chosen Arduino as its favorite tool for DIY constructions and designs.

It is also the most prominent tool used by the young, desktop 3D printer movement. All of the current desktop 3D printers are based on the RepRap project, which is an open-source 3D printer. RepRap uses Arduino as the brains of the machine, so as to communicate with the controlling computer, and in turn make the device perform the requested actions (material extrusion, movement of the object) by controlling its various electronic and electromechanical components.

Again, Arduino was the perfect candidate for this job, due to its low cost, extensibility, and ease of use. As with other examples described earlier and those described further, Arduino is a clear example of how it can be used to democratize technology. To put this into perspective, desktop 3D printing is considered the second industrial revolution, and it is expected to change the way we act, buy, and work by giving us the ability to essentially "print" virtually all the objects and tools we use in our everyday lives (Fig. 7.12).

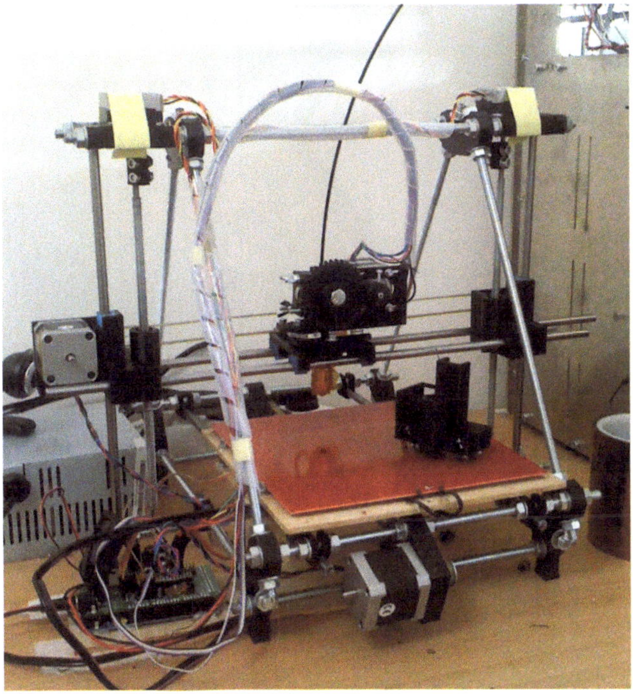

Fig. 7.12 Arduino-based 3D printer

Another great example is the story of Sebastian Alegria, a 14-year-old Chilean teen who created an earthquake warning system based on Arduino, an earthquake detector, and an Ethernet shield (Fig. 7.13).

After living through an earthquake in 2010, Sebastian Alegria constructed this device based on the idea of Japan's earthquake warning system. He used an off-the-shelf earthquake detector, which costs less than $ 100, connected it to an Arduino, giving it the ability to tweet earthquake warnings.

7.4 Arduino in Education

All the aforementioned features made the Arduino the best thing for teaching electronics, interactive designs, or even computer network protocols and applications. To begin with, dozens of schools (elementary or high schools) have introduced Arduino or other compatible devices in their curriculum courses on electronics. On top of that, as Arduino based its success on the Internet and its acceptance there, numerous online courses and tutorials exist and help people learn how to build practically everything from simple digital thermometers to home automation systems.

Fig. 7.13 A 14-year-old Chilean teen who created an earthquake warning system based on Arduino

Fig. 7.14 The Arno board

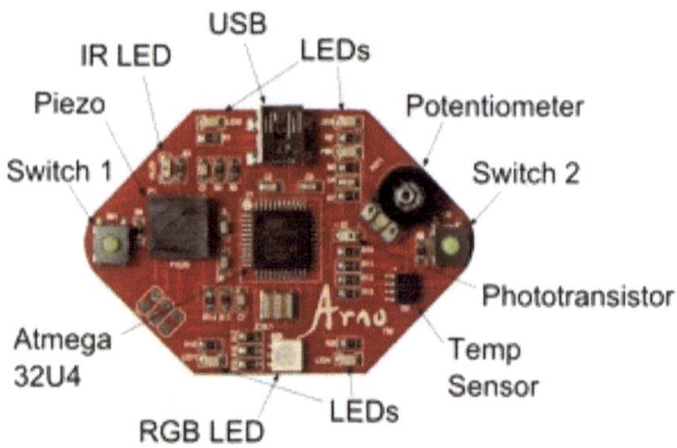

One of the best examples on how Arduino can be used in education is the Arno project (Fig. 7.14). Arno is an Arduino-based device with numerous built-in sensors and actuators that helps young students learn programming for connecting sensor and actuators properly without any hassle. The Arno platform offers a wide range of application examples that help young people understand basic and advanced computer science concepts like sampling and sequencing, as well as the operation of different hardware tools like potentiometers, LEDs, sensors, microphones, and resistors. The most important part is that having everything from the hardware side available and connected to the same chip helps people from being discouraged from starting their journey in the world of electronics by common beginner mistakes and difficulties.

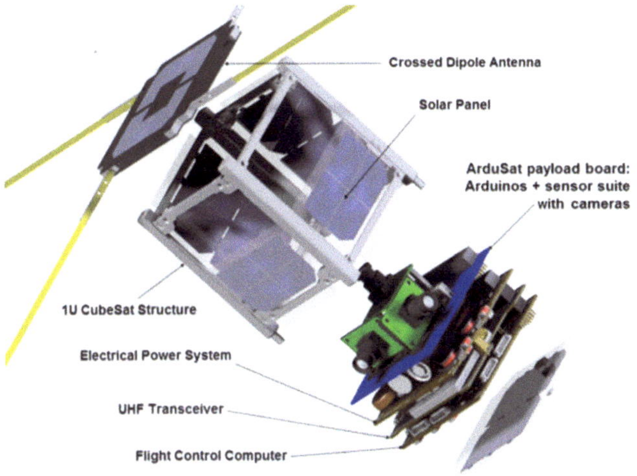

Fig. 7.15 The ArduSat microsatellite

7.5 Arduino in Science

Thanks to its low price, ease of use, and extensibility, Arduino has become one of the favorite tools in science. Arduino is perfect because it supports a big number of sensors, due to which scientists can build their custom measuring devices in days instead of weeks, and with a significantly reduced cost. There is an increasing number of scientific publications and data-gathering applications which use Arduino as the main controller. In fact, Arduino has made possible a number of scientific applications that would otherwise be too hard, or too expensive to use, and it is the most important tool used in democratizing the sciences. One such great example is the ArduSat.

ArduSat is a 10 × 10 × 10 cm microsatellite based on the CubeSat standard, which will be launched in July of 2013 by a company called NanoSatisfy (Fig. 7.15). It features 15 scientific sensors (such as a general-purpose camera, a spectrometer, a Geiger counter, etc.) and Arduino as the controller. The purpose of the project is to allow everyone to rent the satellite for an affordable price (around $ 200 per week), and run their own experiments in space. In essence, it democratizes access to space exploration and space science, which used to be extremely costly, and limited to a handful of scientists who were part of science teams of a space agency like NASA.

7.6 Codebender.cc

Codebender (http://codebender.cc; Figs. 7.16 and 7.17) is a web platform designed to remove the pain of installing and configuring all the tools required to work with Arduino and other platforms. Codebender is a browser-based integrated development environment (IDE) as well as a collaboration hub where users can communicate and

Fig. 7.16 http://codebender.cc

collaborate with each other with a few simple clicks. Its nature as a cross-platform, browser-based tool makes it suitable for people of all ages and expertise offering both advanced tools and operations and simple tutorials and step-by-step guides for people interested in getting familiar with electronics and Arduino.

Amongst other features, codebender provides:

- A full text editor (called Ace) with advanced features,
- The ability to compile the code written using a remote compilation service that contains all available libraries' updates,
- The ability to directly deploy your code to any Arduino-compatible device, simply by installing a browser plugin from most common web browsers (including Internet Explorer, Google Chrome, Mozilla Firefox, Apple Safari, and Opera),
- Community features like code sharing, collaboration, and exchange of ideas.

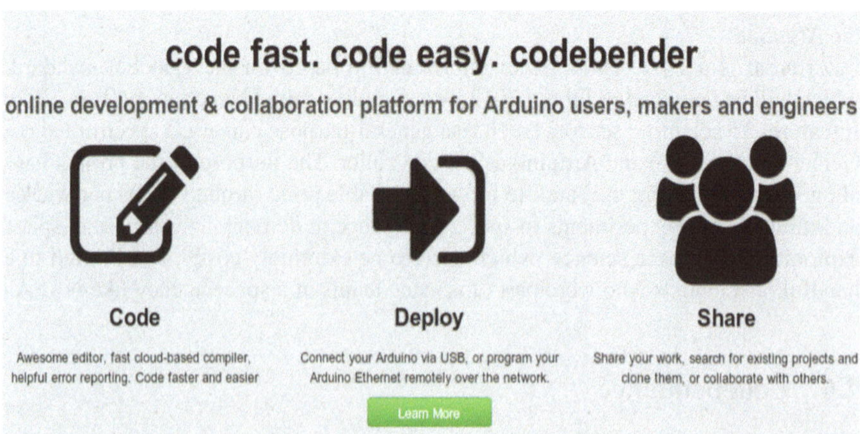

Fig. 7.17 Code fast, code easy, codebender

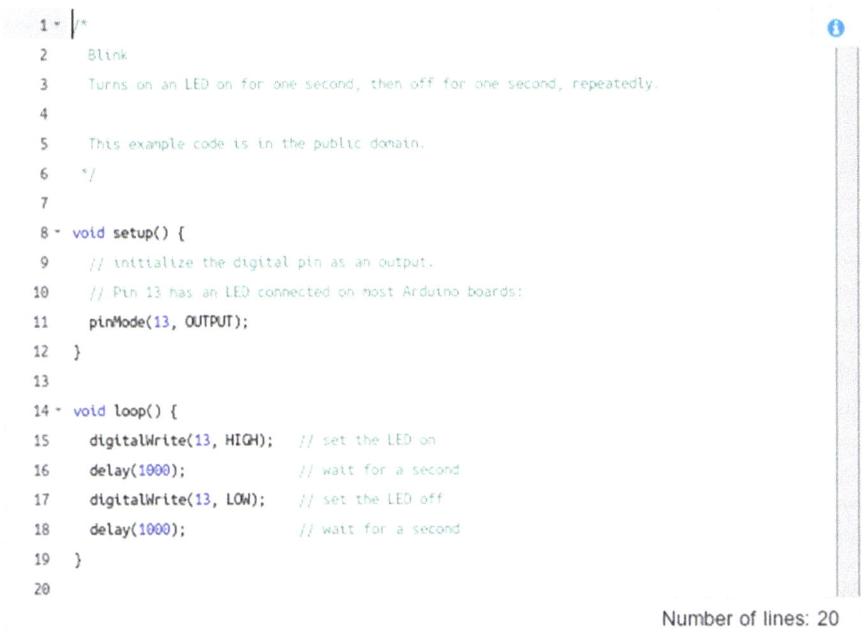

Fig. 7.18 The codebender editor view

7.6.1 Advanced Code Editor

The codebender editor page comes from our personal need to use features available in all other IDEs that were missing for so long from the Arduino and embedded development community (Fig. 7.18). It offers features like proper syntax highlighting, auto-completion, and collapsible code blocks. Based on the powerful, open-source, and well supported Ace editor, it allows users to code faster and more efficiently without worrying about typos and function prototypes as all those come as suggestions and information anywhere they place their cursor.

7.6.2 Remote Compilation Service

The second obstacle for new users is usually the installation of the required toolchain needed for the compilation of their code as well as the maintenance of libraries and updates that can sometimes break dependencies and require days of work to fix. Codebender manages all those so that users do not need to bother with such tedious work. All libraries are available via a library management service for users to work with them without any extra steps. Also, updates are managed in a single point and tested thoroughly before their deployment to the system. Additionally, codebender

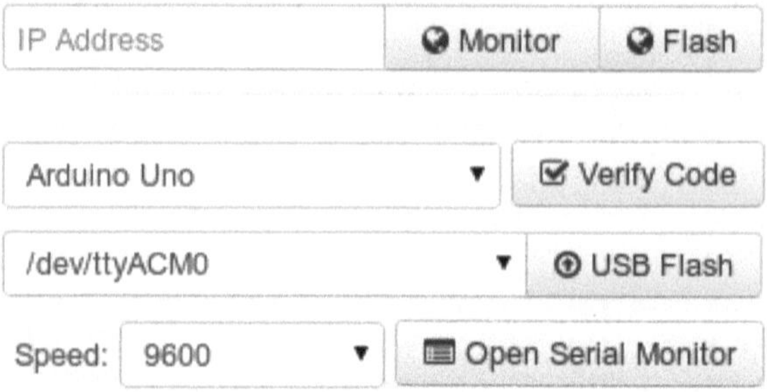

Fig. 7.19 The codebender browser plugin tools

offers advanced code analysis using Clang, an advanced compiler that provides easy to read and understand messages for errors or mistypes.

7.7 Easy Deployment Right from the Browser

When you are done writing your code, what you need is to deploy it to your Arduino as fast as possible to watch it do the magic stuff you want it to. Codebender offers you the ability to do this straight from your browser without any more pain. In fact, codebender offers two different solutions based on the nature of your project: either deploying your application to an Arduino connected to your computer via a USB cable, or sending it to an Arduino connected to the Internet on the other side of the world. The first solution is made possible using the codebender browser plugin (Fig. 7.19). This plugin offers the ability to communicate with any supported device with a single click and no extra configuration. The second option comes to help more advanced users who need to easily update the firmware on an existing project without having to dismantle it every time to bring it to their computer (i.e., an Ethernet weather station on the roof). The code is simply uploaded to the Arduino using the simple trivial file transfer protocol (TFTP) protocol as part of the Arduino's bootloader. The bootloader is a small but very important part of code that runs whenever the device is powered on and checks for a reprogram request. In essence, it is for embedded processors what BIOS is to PCs.

Moreover, codebender offers the ability for two-way communication with the Arduino using either the USB or a simple network or Internet connection. The USB communication is achieved via the plugin, using serial communication. Over the network, users can enter their Arduino's IP address to send and receive messages via a WebSocket connection, without the need for routing changes in the home LAN or enterprise network.

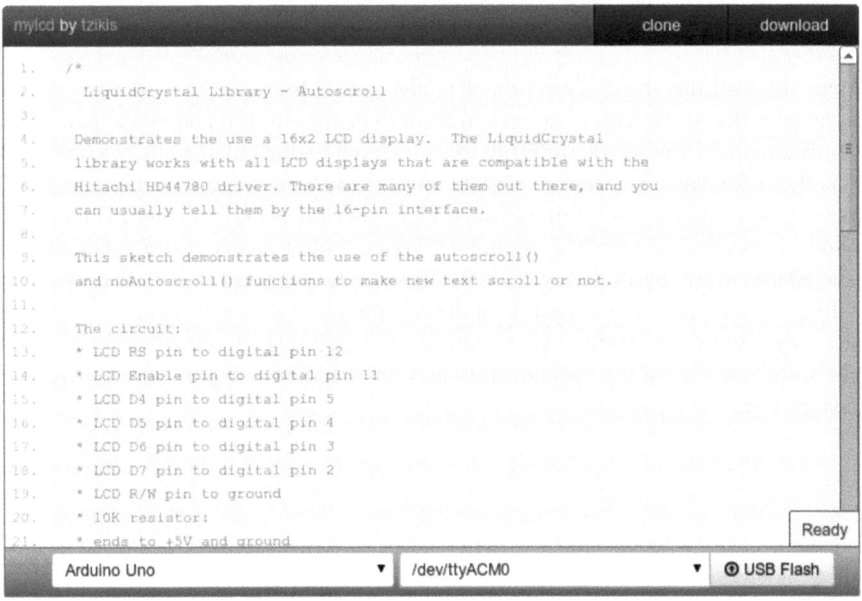

Fig. 7.20 The codebender embeddable view

7.7.1 Community Features

Another great feature provided by codebender, is the embeddable project view (Fig. 7.20). This function allows users to easily present their work on the website or blog by simply copy pasting a simple piece of code. Apart from showcasing your work in a very good looking way, the embeddable view offers viewers the ability to flash the showcased code to their device directly without any trouble at all. Moreover, as the code is actually stored on the server, you do not need to worry about updating the embeddable view whenever you make the simplest change in the project. The code will always be up-to-date right after you hit the save button in the editor.

Codebender also offer various community features that allow users develop relations and help each other to make the online world a better place. Features like Karma and Points that users gain based on their contribution to the community (e.g., when they clone projects, help resolve a bug, share their work, or discuss ideas and projects).

Cloning projects is also another way for the community to bond. Users can find ideas that are pretty close to what they want to build and learn from other people's experiences and expertise. Codebender also offers a nice way for people to communicate over a project or a part of it using a commentary mechanism that provides easy discussions over lines of code functions or files.

7.8 Advanced Features

Apart from all the above, codebender is also a tool for more advanced users and companies that work with Arduino and want to benefit from it and make their work a little simpler. Projects and information are stored on the cloud so there is no need to bother with backups, synchronization, or send out emails to colleagues about bug fixes or suggestions. Everything is done from your browser, from any PC (running Windows, Linux or MacOs), or mobile device. Additionally, extra functionality is available to the users that require it, like private projects (as not all people are willing to share their work right from the beginning) or personal boards to test new prototypes and functionality. Also, there is the possibility of collaboration features like simultaneous editing, organization accounts, and auto-deployment to Arduino devices right from your browser.

Chapter 8
The Internet of Things: How WSNs Fit Into The Picture

Aggeliki Pragiati

Abstract The Internet of today offers access to content through the web via multiple channels. The next evolution will make it possible to access information related to our physical environment through a generalized connectivity of everyday objects. A car may be able to report the status of its various subsystems using embedded communicating sensors for remote diagnosis; personal devices may deliver the latest status of healthcare information of remotely cared patients to a central spot; environmental data may be collected and processed globally for real time decision making. Access to information relating to our surrounding environment is made possible through communicating objects able to interact with that environment and react to events. This evolution of networked devices is also known as the Internet of Things (IoT) and creates space for new application classes. IoT is a highly promising economic sector for sustainability, growth, and innovation. The challenge is to assess the right abstraction and complexity level of "things" involved in order to promote research on the benefits and the control that we want to retain over an environment where machines will gather, exchange, process, and store information automatically.

8.1 Introduction

Access to information relating to our surrounding environment is made possible through communicating objects able to interact with that environment and react to events. This evolution of networked devices is also known as the Internet of Things (IoT), invading applications in a growing number of fields, the need for fast yet accurate exploration of optimal alternatives, thus necessitating the development of structured and efficient performance evaluation strategies.

Complex and demanding applications are more and more associated with the application of wireless sensor network (WSN) technology. Starting from the lower level of communication algorithms to the higher application level and its associated functionality requirements and constraints, WSNs are used in a variety of application domains such as environment, health, security, military, or urban. Each scenario may require collaborative sensing, communication, and computation among multiple

A. Pragiati (✉)
Hellenic Telecommunications & Post Commission (EETT), Athens, Greece
e-mail: apragiati@eett.gr

N. Sklavos et al. (eds.), *System-Level Design Methodologies for Telecommunication*,
DOI 10.1007/978-3-319-00663-5_8,

sensors that observe moving objects, physical effects, and/or environmental events and it is commonly structured in tasks, such as deployment, application functionality, and information exchange.

Meeting the application requirements could greatly depend on optimal and energy-efficient nodes placement. The actual deployment affects network properties such as node density and topology but may also predetermine the data collection and routing mechanisms by providing connectivity degree and sensing coverage. Prudent planning and analysis of different deployment strategies could lead to network efficiency with respect to energy, cost, speed, and lifetime.

From the network point of view, there is a variety of protocols trying to enhance the performance of the network. Still no standard one has been established. Metrics used in route selection, such as power awareness and disconnection management, are issues that still need a lot of research. However, a good routing strategy requires an efficient underlying Medium Access Protocol (MAC) to support network performance. Reliable and efficient sharing of the wireless transmission medium, scalability, and mobility are critical issues when designing a network protocol, introducing design problems difficult to overcome. Cross-layer optimization intends to improve the existing approach that a layer in isolation does not lead to efficiency, since it ignores critical interactions and correlations that should be exploited. Last but not least, security is a challenging demand in complex data-intensive WSN applications. Ensuring data confidentiality, integrity, and authentication are some issues in WSN communication security.

As models for mobile wireless networking become more popular, their appeal comes from the fact that they can operate autonomously without the need for existing infrastructure. This great benefit can be seen even more clearly when looking at the many problems that the use of WSN technology solves. The applications of WSN technology have been classified into four main categories: environmental monitoring, healthcare, security, and additional applications. At most of these cases, an efficient localization algorithm is expected to predict the path of the mobile object with high accuracy. Localization applications using WSNs range from health monitoring to house safety and even to military surveillance. On the other hand, the system performance is affected by a series of parameters that need to be addressed and co-evaluated during the simulation process, which is not always possible. These parameters are related to the communication conditions, to the mobility pattern, to the object's positions, and other application characteristics. The mobile object speed and mobility pattern are important parameters as they directly affect the object's communication range, indirectly its sensing capabilities, and thus application performance. Moreover, the network topology may facilitate or harden the localization task, i.e., a randomly deployed set of nodes has less probability to achieve high communication coverage in the localization area than a predefined deployment scheme. Finally, the limited resources of WSNs impose restrictions to their capabilities and performance; the limited memory restricts the use of complicated and demanding algorithms and the limited battery storage restricts the power consumption and extensive execution time. After sensing and processing data, WSN nodes need to consume it or send it immediately to a storage point.

In this chapter, trends and characteristics of IoT are discussed, with particular focus on the application fields and WSN particularities. Section 2 presents the most common features of IoT, while Sect. 3 attempts to give an overview of existing and future application clusters for IoT. As the underlying network plays the role of the glue to put it all together, Sect. 4 discusses WSN deployment strategies and Sect. 5 presents a short communication mechanism overview. Section 6 exhibits framework architectures and approaches that support the interaction between "things", while Sect. 7 introduces general modeling aspects for some of IoT frameworks. Finally, conclusions are drawn in Sect. 8.

8.2 Trends and Characteristics

Architecture	Most likely, the system architecture is event-driven [1] and designed following a *bottom-up approach*. Model-driven and functional approaches are combined in agent-based systems apt to treat exceptions and unusual processes. Since events are based on the context of the event itself, common standards cannot fully address every context or use.
Time	Because of the huge number of parallel and simultaneous events, time cannot be treated as a common and linear dimension but will depend on each *entity* (object, process, information system, etc.). This IoT is accordingly based on massive parallel IT systems (parallel computing).
Intelligence	Even though ambient intelligence does not necessarily require interconnected structures, a trend to integrate the concepts of the IoT and autonomous control is emerging. The IoT may be viewed as a nondeterministic and open network, in which auto-organized, intelligent entities, and virtual objects are interoperable yet independent, depending on the context. Embedded intelligence [2] presents an "AI-oriented" perspective of IoT, which can be more clearly defined as: leveraging the capacity to collect and analyze the digital traces left by people when interacting with widely deployed smart things to discover the knowledge about human life, environment interaction, as well as social connection/behavior.
Size	The IoT would cover trillions of objects, and be able to follow the movement of those trackable objects [3].
Complexity	Any modern system is characterized and treated as a complex one, due to the huge number of different links and interactions among autonomous objects.
Space considerations	In an IoT, the precise geographic location of a thing is critical [4]. As the Internet has been primarily used to manage information processed by people, an object location in time and space, have been less critical to track. However, emerging applications demand

that objects be self-organized and connected by location. Important constraints for such an achievement are the variable spatial scales, the great amounts of data, and a neighbor awareness mechanism. In order to support the self-X prerequisite for the IoT, objects should be able to act on their own time-space context.

8.3 Applications

Among the widely-used application fields for the IoT, environmental applications play a key role. The use of WSNs in such applications can contribute to the development of hazard response systems, evolving into "intelligent" sensor networks. These networks comprise nodes and communication systems, which actively transmit their data to a Sensor Network Server (SNS) where this data can be integrated with other environmental datasets. These sensor nodes can be fixed or mobile and range in scale depending on the environment being sensed. Large-scale single function networks tend to use large single purpose nodes to cover a wide geographical area. Localized Multifunction Sensor Networks typically monitor in detail a small area, often with wireless ad hoc systems. Environmental Sensor Networks provide a 'virtual' connection with the environment.

A classification of potential application scenarios was performed taking into account the possible implementation of the wireless sensor network in the environmental domain: Meteorological monitoring, geological monitoring, habitat monitoring, pollution monitoring, and energy monitoring.

Geological Monitoring Geological monitoring refers to the control, supervision, and study of several physical geological magnitudes, to enhance the understanding of the earth's state. Geological stations can provide information on underground temperature, water content, seismic activity, tilt, and displacement. These measurements are useful to forecast and predict harsh natural phenomena, namely catastrophe prediction. The common factor of most geological disasters as such is the fact that they are related to underground activity [5]. As WSN devices are fully functional below ground, they provide the best choice to form the IoT for geological monitoring.

Habitat Monitoring Habitat refers to the biophysical medium occupied continuously, periodically, or occasionally by an organism or group of organisms, for which it detects tendencies of abundance of species. Two major application clusters are identified, namely wild natural scenarios for both animal and vegetation, and improving production in agriculture, livestock farming, or fish farming. Sensor networks in such applications provide:

- Long-term data collection at scales difficult to obtain otherwise.
- Localized measurements and detailed information hard to obtain through traditional instrumentation.
- Easy interaction among the in situ network (WSN) and other external networks.

Pollution Monitoring Air pollution, water pollution, noise pollution, and radioactive contamination to name a few, all need to be controlled and monitored in the attempt to reduce their damaging consequences. The main forms of pollution include:

- Air pollution
- Water pollution
- Noise pollution
- Other pollution i.e., light pollution, visual pollution, and radioactive contamination.

Application scenarios are classified in this field as: Air pollution monitoring, water pollution monitoring, and acoustic pollution monitoring and waste management. Typical systems centralize information coming from sensors monitoring air pollution.

Energy Monitoring Any kind of economic activity requires energy resources, which in their turn contribute to air and water pollution. While burning fossil fuels produces instantaneous supply of electricity, it also generates air pollutants including carbon dioxide, sulfur dioxide, and nitrogen oxides. Since energy cost has become a significant economic and environmental factor, energy resources management has evolved to a crucial dimension of global ecosystem. Energy management involves utilizing the available energy resources more effectively, while at the same time eliminating energy. The exploitation of WSNs aims at energy consumption reduction and thus, energy conservation. Energy monitoring applications are classified according to industrial or domestic environments.

The WSN technology could potentially impact a number of healthcare applications, such as medical treatment, pre- and post-hospital patient monitoring, people rescue, and early disease warning systems, in improving not only the quality of life, but also in benefiting society as a whole. Health applications are classified into five separate categories: patient monitoring, disability assistance, people rescue, bio-surveillance, and smart surrounding.

Patient Monitoring observes the patient health state either in the hospital or at a home environment. Current systems that are used for long-term patient monitoring require the use of wires, whereas in WSNs this is not necessary. The measurements of patients' vital signs can be useful not only for medical records and treatments, but also for later rehabilitations. Patient monitoring in hospital environments aims at continuously collecting patients' vital signs, allowing doctors and nurses to continuously monitor their status and to react to changes [6, 7].

Disability Assistance refers to scenarios, where smart sensors operate within the human body to counteract organ weaknesses or monitor important physiological parameters or particular organ viability. For the treatment of debilitating neurological disorders, implantable, multi-programmable brain stimulators have already been developed, saving the patient from surgical operations. In cardiology, the value of the implantable cardioverter-defibrillator has increasingly been recognized for the effective prevention of sudden cardiac death [8].

People Rescue In emergency or disaster scenarios, if people are outfitted with tiny wireless badges, the rescue teams and medics will be guided much faster to the victims, allowing large numbers of casualties to be prevented [9]. These sensors would relay continuous data to nearby paramedics and emergency medical technicians, who would use mobile personal digital assistants (PDAs) or mobile personal computer (PC)-based systems in ambulances to capture all vital patient data. They could thus monitor and care for several patients at once and be alerted to any changes in the patient's physiologic status. The information network includes communication with the rescue teams as well as communication with the hospitals' information system, allowing for better coordination between the emergency rescue teams and the hospitals with the facilities and resources to care for patients in critical condition. Patients in accidents can greatly benefit from technologies that continuously monitor their vital status and track their locations until they are admitted to the hospital. The sensors support for functions of vital sign monitoring, location tracking, medical record storage, and triage status tracking. These sensors would continuously relay data to nearby paramedics and emergency medical technicians, who would use mobile PDAs to capture all vital data. Thus, they could monitor and care for several patients at once and be alerted to any changes in the patient's physiologic status.

Bio-surveillance assists public health experts in computing the likelihood of a deadly disease outbreak among the human population [9]. A series of sensors can collect and examine samples from the air, soil, and water and use weather conditions to predict the epidemiological spreading of the disease [10, 11]. This prediction allows for fast and effective reaction providing fast emergency response, medical care, and consequence management needs.

Smart Surrounding solves important social problems such as caretaking for the chronically ill, elderly people, and people with mental and physical disabilities. Such services of health care are provided to citizens at home, allowing them to have a normal life [11, 12].

In the security domain, multiple types of functionality coexist, such as target tracking and localization, detection of toxic chemicals, rescue, and homeland security. However, what makes this domain particular is that this complexity increases by the large network scale, by the multi-hop ad hoc network of tiny resources and energy-constrained sensor nodes.

European research in WSNs focuses on civilian applications, on a smaller scale, heterogeneous hardware, single-hop networks. However, major breakthroughs have been achieved in the military application field. Surveillance is taken as the process of monitoring people's behavior, objects or processes within systems, for security or social control. Surveillance is mainly divided in two categories, namely indoor and outdoor surveillance.

Indoor Surveillance is applied by surveillance systems, placed in a private environment, i.e., in a home and those located in public buildings such as hospitals, museums, and airports, among others. The objects may have, different energy capacities, processing capabilities, positions, and radio coverage. Such WSNs are easy to deploy, have ubiquitous connection, are low maintenance and unobtrusive, saving costs in wiring installation. Some sensors, such as thermal sensors, integrate processing units that estimate the alarm conditions and presence controlling can be carried out with volumetric sensors. These sensors receive infrared radiations from elements, or generate an invisible wide detection field. Volumetric devices are based on pyroelectric sensors, which convert infrared radiation variations emitted by elements into small current, allowing the sensor to examine infrared radiations in a fixed area of the room installed. Video surveillance systems are necessary to identify the alarm source detected by volumetric devices, and small amplitude and speed variation at any given time triggers an alarm. Combination of devices is also possible. Movement video detection systems transform the video capturing capabilities of a closed television circuit into an image capturing detection system, analyzing video output to generate an evaluation field.

Outdoor Surveillance usually provides perimeter security whether to keep objects inside the perimeter or to keep intruders out of a certain area [13]. When using invisible surveillance, it is fundamental that intruders are not able to detect its presence. Such solutions include:

- *Magnetic detection system* based on the magnetic anomalies passive detection, which allows the detection of any intruder that carries ferromagnetic metal objects.
- *Vibration detection system with* sensors are attached to a wire fence, for detecting any vibration produced by climbing or wire-cutting of the enclosure [14, 15].
- *Electromagnetic field detection fixed system* makes use of a fixed set of volumetric sensors for perimeter intruder detection, which generate an electromagnetic field around two wires, both field-emitter and receiver, buried in the ground.
- *Electromagnetic field detection portable system* allowing for fast installation and easy handling for perimeter security.
- *Microwave perimeter control* establishing a microwave barrier between sensors, whose status is analyzed by a digital signal processor.

These systems are based on amplitude and phase disturbance detection, emitting the signal by a small antenna. When intrusion activity patterns are received in the perimeter antenna, the system activates an alarm. These systems can be combined with WSN technologies as each antenna forms an isolated node.

Airport security provides a first line of defense by attempting to stop potential attackers from bringing weapons or bombs into the airport. If airport security is successful then the chances of these devices getting onto aircrafts are greatly reduced. Bomb detection, luggage tracking, drug detection, and hijacking are all current threats in many airports throughout the world, which is why the advancements in WSN technology are so important today.

The benefits of WSNs in military applications can be seen in homeland security, military vehicle operation and maintenance, and battlefield monitoring. Active and passive sensors can be used to detect the presence of nuclear, biological, and chemical agents. Passive sensors detect a change in the natural energy field caused or emitted by a target, based on capacitance, heat, sound, and vibration. Active sensors transmit energy and detect a change in the received energy as the target comes within range. By linking wireless routers to video cameras, images can be transmitted from disaster areas to vehicles, fires stations, command centers, and other public safety agencies. Officers in command centers can view the video of an incident, analyze footage, and relay orders to a response team, all in real time.

In addition to the aforementioned applications, the list is continuously spreading, including areas like structural health monitoring, building monitoring, building control, automotive monitoring, traffic monitoring, industrial process control, and asset and warehouse monitoring.

Structural Health Monitoring Life cycle monitoring of civil infrastructures such as bridges and buildings is critical to long-term operational cost and the safety of aging structures. Events like earthquakes can cause enormous damage to civil infrastructures without producing any apparent visible damage, causing life threatening conditions in the structure. Near real-time structural monitoring of civil infrastructure reduces the loss of human lives by warning for hazardous structures and impending collapses. As far as bridge structure monitoring is concerned, the following techniques are quite popular:

- *Slow monitoring* measures slow phenomena like temperature changes, settling and concrete relaxing, therefore the sampling frequency can be hours. Variables measured could be air temperature, straining in several axes, steel distortion, and solar radiation, in places like boards and pillars.
- *Fast monitoring* measures fast phenomena like traffic, wind, and earthquake effects. Sampling frequency may be variable, seconds or milliseconds, with levels depending on variable speed changes and the sensors for these measurements are accelerometers, wind speed, and direction meters.
- *Corrosion monitoring* is designed to detect steel corrosion and thus has a slow measurement interval of perhaps days.

Building Monitoring and Control embed sensors in buildings to reduce energy costs by monitoring the temperature and lighting conditions. The information obtained is then used to regulate heating systems, cooling systems, ventilators, lights, and computer servers [15]. Sensors in a ventilation system may also be able to detect biological agents or chemical pollutants.

Automotive Monitoring, as well as Traffic and Transport Monitoring use magnetic field measurements. The sensor module detects passing vehicles by measuring disturbances in the Earth's magnetic field, caused by passing vehicles. Almost all of today's road vehicles, even vehicles with polymer body panels, contain a large mass of steel. The steel has a much higher magnetic permeability than the surrounding air, which concentrates the flux lines of the Earth's magnetic field, increasing the magnitude of the B-field inside and in the immediate vicinity of the vehicle. This disturbance is detectable as far away as 15 m from the vehicle.

Traffic Monitoring has grown to be mandatory in order to handle the increasing congestion in public road networks of many countries [16]. Any remedial strategy for efficient road management requires measurement of traffic conditions either to optimize traffic signal settings based on traffic queue lengths or to better plan a driver's. Most conventional traffic surveillance systems use intrusive sensors, including inductive loop detectors, micro-loop probes, and pneumatic road tubes, because of their high accuracy for vehicle detection. However, these sensors usually require a high installation and maintenance cost. Heterogeneous network topologies of WSNs and access points interoperate to get traffic information generated by sensor nodes and transfer it to central points for control.

Industrial Process Control needs real-time access to information regarding the environment of the industrial plant, processes, and equipment to prevent disruption [17]. WSNs offer lower system, infrastructure, and operating costs, as well as improvement of product quality, streamlining of operations, easier upgrading, greater physical mobility, and more freedom. Unlike traditional wired networks, the sensors of a WSN can be deployed in the bearings of motors, oil pumps, whirring engines, packing crates, and many other unpleasant, inaccessible, or hazardous environments that are inaccessible with wired systems [15, 18]. Smart sensors allow for monitoring the equipment status in the field and preventing imminent failures. Condition-based monitoring has significantly reduced the cost of service and maintenance, increased the machines lifetime, and even saved lives.

Asset and Warehouse Monitoring WSNs are used to monitor and track assets such as lorries or other equipment, especially in an area without a fixed networking infrastructure [15], for industries such as oil, gas, and aerospace. Tracking sensors vary from GPS-equipped locators to passive radio-frequency identification (RFID) tags and the automated logging system reduces errors of manual data entry. Industries can significantly improve asset utilization using real-time information about equipment location and condition. Furthermore, the asset information can be linked to other systems such as enterprise resource planning (ERP).

Apart from monitoring, most of the application classes presented above also require tracking or localization functionality. As many localization techniques exist in the literature, the most widely used for fixed network topologies are discussed and analyzed with respect to the network and mobility parameters to select the most representative one for studying localization performance of a mobile target in fixed WSN topologies.

At the initial phase of the localization algorithm, every fixed node makes known its position to the rest of the nodes in the network. At the end of this phase, nodes keep only their neighbor's positions so as not to overload their memory. Once a mobile target enters the localization area, the fixed nodes attempt to track its path based on their relative distances from the target. The result of the localization algorithm is then either communicated to a node centrally placed in the network area or consumed by the mobile target or even by a set of fixed nodes in the network. This last step depends on the application case.

The complexity of impact factors and varying application cases affecting the localization task success have led to a wide variety of existing localization techniques targeted to specific application requirements. These techniques differ in the type of signal used to estimate the target position, with the most popular being the radio frequency (RF) signal, ultrasound, infrared, and the received signal strength (RSS). Some techniques have been developed for fixed topologies and others for dynamically changing networks. Focusing on the fixed topology localization techniques, grid, and random scenarios are discriminated as the varying impact factors are seriously affecting localization performance.

For grid topologies, the most common technique is the fingerprinting approach, which computes the target position based on the comparison between the RSS indication (RSSI) and predefined signal strength measurements stored in a database [19]. For random topologies, the most famous technique is hop-counting, which uses RF signals and hop count tables for every node's neighbors [20].

8.4 Wireless Sensor Network Deployment Characteristics

WSN deployment considers the features and requirements of the sensor network among included in taxonomy. WSN particularities carry characteristics that range from the application to the node hardware abstraction and include energy resources and power consumption, processing capabilities, network topology, or communication protocol among others. The objective of WSN deployment is to achieve the required sensing coverage of the wireless sensor network with an optimal number of nodes, while complying with several constraints as power consumption, reliability, cost, scalability, latency, etc., and to propose an adequate network topology and routing for the information dissemination.

Depending on the accessibility of the application scenario, a first classification can be made dividing the deployment process in two groups: manual and random placement. This simple classification allows defining different parameters in order to guarantee the correct working of the network for the specific application. Then, a real world deployment should be carried out, followed by debugging and test operations. The high complexity and size of WSN systems renders testing and network deployment of thousands of nodes increasingly hard, for which simulation can rarely provide insight. The following factors play a significant role on network deployment and thus, on WSN behavior.

Network size is suggested by the area and the number of nodes. This is a very important requirement for the communication protocol efficiency. The nodes mobility degree is closely coupled with the size as well. More precisely, setting the boundary among large, medium, and small size is not a trivial question. Network density, defined by nodes per area unit, is a more adequate term for describing the network capabilities with reference to communication coverage. In this context, more or less nodes would be required per square meter, depending on the application parameters to measure, the type of application i.e., tracking application may require more nodes to support the localization algorithms, or the type of sensors and sensing range.

Coming to nodes mobility, some applications require that the nodes move through the environment under measurement. Even in cases of a manual initial deployment, the network topology may change due to nodes movements. In these cases, it could be possible to go through a second stage of deployment, in which the network controls itself. The nodes are able to detect the positions of other nodes, and change their positions accordingly, maintaining the features to achieve suitable quality of service.

The lifetime is a very restricting requirement, because WSN nodes must be small in order to make the network as invisible as possible to the environment, in which the measures are going to be taken. The small size requirement comes at the price of very limited energy resources, so it is very important to take into account this parameter. For example, the energy consumed for communication in wireless links is determined greatly by the distance between the nodes, and a trade-off must be applied among nodes distances and energy consumption.

The dynamic nature of WSN applications, supporting the Internet of things (IoT), imposes that new nodes could join the network at any point in time after the initial WSN deployment, either due to node replacement or because new parameters have to be measured. Scalability has influence on the communication protocol and on network topology.

Depending on the kind of the physical parameters that the network has to measure and on the existence or not of an infrastructure, the nodes that compound the WSN are in most cases different in hardware and software. This heterogeneity leads to different deployment decisions. On the other side, the existence of infrastructure in the network means that apart from ad hoc communication, other resources may be used as well in the communication process, i.e., Wi-Fi access points, GPS, etc.

Quality of Service (QoS) is probably the most important requirement because it makes reference to the system behavior and performance. This requirement is strongly translated to other network features as security, coverage, and fault tolerance.

The applications require specific sensors to measure environmental parameters. In this context, the type of sensors may affect or relate to some of the above mentioned parameters. Depending on the application, these nodes can be selected from a set of available nodes in the market, developed ad hoc, or imposed by the client. In this situation, some of the requirements for the communication protocol, radio range, energy resources, processing capabilities, etc. are imposed by the hardware platform.

8.5 Wireless Sensor Network Communication Aspects

Three major requirements are the target of every network protocol design: bounded delay, power awareness, and low overhead imposed to the network. These three requirements however seem to be contradictory to each other, thus demanding a trade-off in order to enhance one aspect of network performance at the expense of the rest.

8.5.1 Network Layer

Routing protocols are basically grouped by the proactive or reactive way they create routes. Both approaches can be applied in WSNs depending on the application data creation pattern, establishing one or multiple routes. In case of multiple routes, the protocol selects one of them based on a link-route metric. Some approaches relate this metric to power [21] and some others to the real-time performance of the network either indirectly by hop-count or directly by selecting the route that formed fastest after the initial route creation request [22]. However, no standard metric exists, since each application imposes different requirements on the routing protocol.

Hierarchical protocols like Leach [23] try to minimize protocol overhead through localization of data transmission using clusters and cluster heads. Leach includes distributed cluster formation, local processing to reduce global communication, and randomized rotation of the cluster-heads to minimize the possibility of premature energy exhaustion of nodes having this role. Although it is a promising routing protocol, it is not always suitable, since the mechanism to elect cluster-heads imposes overhead to the network and local processing cannot be used in cases of tracking, where data aggregation is needed. The only functionality provided is the grouping of data of several packets to one, in order to minimize the packets sent to the WSN sink. In this way, the possibility of collisions in a contention based MAC is lowered, but larger packets are transmitted degrading network performance.

SPIN [24] is a family of protocols used to efficiently disseminate information in a WSN using data negotiation and resource-adaptive algorithms. Nodes running SPIN assign a high-level name to data, called meta-data, and perform meta-data negotiations before any data transmission, assuring that there is no redundant data sent throughout the network. In addition, SPIN is adapted based on the remaining energy and uses data aggregation, which is dependent on the application requirements and nature. For health data-intense application, data is deterministic in nature, providing data that cannot be aggregated without losing vital information.

PEGASIS is a greedy chain protocol that is promising for data-gathering problems in WSNs. Nodes take turns to transmit the fused data to the base station (BS) to balance the energy depletion in the network, while preserving its robustness as they die at random locations. Distributing the energy load among the nodes increases the lifetime and quality of the network. PEGASIS uses controlled transmission power in order for nodes to be able to alternatively transmit to the BS but implying long transmission range which most of the times is not the case for WSN platforms. Even more, data fusion is not applicable to health applications that demand localization because of possible loss of vital information, and mobility of nodes is not supported, which is important for other applications too.

Another interesting routing approach is FloodNet Adaptive Routing (FAR) [22], designed for use in the FloodNet project. FAR examines the impact of diverse reporting rates on protocol design. It incorporates mechanisms for interest diffusion, neighbor status maintenance, a routing algorithm that uses the above mechanisms and a mathematic weight formula to select routes to the WSN sink. The mathematic nature of the route selection metric allows for further fine tuning or partial redesign that could fit best in any application. However, studies made on FAR until now, do not take into consideration mobility or dynamic topology, which are basic characteristics for most of today's WSN applications.

8.5.2 Media Access Control Layer

Sensor network MAC layer protocol has major problems to face as far as resources' waste is concerned [25–27]. Identification and handling of collisions impose serious delay penalties and lead to power consumption overhead. In order for a WSN to be configured, a considerable number of packets not carrying user data must be maintained and managed. These are control packets and minimization of their number leads to power conservation and reduction of unnecessary workload. In addition, a considerable percentage of power consumption of a node is due to overhearing when receiving and processing a packet not intended for the specific node. Finally, idle listening is also a major problem of a MAC protocol.

In order to face these problems, MAC protocols follow certain techniques, with carrier sense multiple access (CSMA) being quite popular, utilized by SMAC [27] and BMAC [25]. CSMA algorithms are decentralized, without any control needed from a single entity, which is more suited to the distributed nature of a WSN. Of course, in this case no bounded delay is guaranteed, but if the network data load is kept within limits, then average delays will be low, even when collisions and hidden-exposed node problems hinder the network performance.

To face the problem of collisions [23], a typical time division multiple access (TDMA) approach promises deterministic delays at the expense of higher access times and low bandwidth usage in low data traffic state, like in cases of environmental monitoring. However, in health applications with many mobile nodes entering and exiting wireless domains, where vital signs need to be continuously monitored, this is not possible.

Hybrid techniques, like Z-MAC [22], follow the CSMA technique in low traffic conditions, but when traffic increases, it adapts its functionality to TDMA. Finally, CrossMAC [26] follows the cross-layer approach, with control packets containing routing information and considerably facilitating the scheduling of sleep–wake period of each node. There have been many approaches for cross-layer design, among Network, MAC, and PHY layers. Some of the dominant approaches in cross-layer design for WSNs are presented in [28]. Using information from MAC and PHY, routing algorithms can enhance their performance while minimizing the overhead imposed by them. Collision detection information from MAC layer can pinpoint possible broken links resulting in links and routes deletion at the Network layer. Transmission power manipulation in the PHY can be used to control connectivity of each node, resulting in avoiding network partitioning while maintaining a good connectivity degree for each node. All the above techniques are open to evaluation as far as WSN application specifics are concerned with cross-layer design being the most promising.

8.5.3 *Security*

With respect to security provisioning, different techniques exist varying the characteristics and demands of such mechanisms [29, 30]. All security mechanisms are based upon cryptographic algorithms, discriminated in two major categories: private (symmetric) and public (asymmetric). Private cryptography is less demanding on computational power, but this imposes considerable control overhead. Public cryptography solves the key number issue, but it affects negatively the node performance and results into larger cipher data, burdening memory usage.

Efficient management of the keys is crucial for security and the respective key management techniques can be indicated by the chosen cryptographic algorithm. Bootstrapping key management is in accordance with the WSN application nature, since each node shares a key only with the BS and all other keys are derived from this. However, this introduces a single point of failure, which is a very significant disadvantage for WSNs. Key management pre-distribution is very interesting, according to which a subset is chosen from a symmetric-key pool and distributed to each node. In this way, not all nodes can communicate with each other, but by utilizing smart distribution and by exploiting statistical research, full connectivity across the network can be guaranteed. Polynomial-based, Blom's matrix-based, deterministic and pure probabilistic key pre-distribution are modifications of the pre-distribution algorithm, presenting a clear trade-off between efficiency and sensor node needs.

In WSNs, the main trade-off is between security level and energy cost. However for this trade-off, no optimal solution for each application scenario exists. Instead, general guidelines can be given, which combined with the criticality of the application can provide the best solution. Thus, public and private cryptography can be combined by utilizing the former for setting up a private key between the communicating parties, while exploiting the latter characteristics of less computational demands to transfer the actual data. In addition, indicative analysis of security levels can be achieved by varying major cryptographic algorithm parameters [31].

8.6 Frameworks

IoT frameworks might help support the interaction between "things" and allow for more complex structures like distributed computing and distributed applications. Stage of the art of IoT frameworks focus on real-time data logging solutions (Xively), allowing for many "things" to act and interact. Several network services regarding sensor, control, multimedia, and many other applications are becoming more and more popular today. It is expected that every mobile electronic device has a wireless network interface allowing it to join general, heterogeneous networks and to exchange information with other devices. From the user point of view, these devices must work transparently. At the same time, any communication must be performed efficiently with respect to quality of service (QoS) and heterogeneity.

A primary concern of the interoperability framework is to provide seamless cooperation and communication between devices with different characteristics either software or hardware. Different capabilities result in trade-offs that must be taken into consideration to assure optimal network operation and resource utilization. Thus, several mechanisms are required to determine which devices are optimal for which tasks. In that way as users move from one device to another, applications are able to adjust to different capabilities resulting in an overall optimal performance.

Even more since mobility support is mandatory to any wireless device, requirements are more complex. These devices must be smaller in size to allow users to conveniently move around with them; this size constraint limits other resources such as screen size, processing power, and battery life that have major impact on connectivity and the development of services and applications.

Nevertheless, device-level interoperability must be extended to technology and protocol interoperability as well. Nowadays many different technologies can provide efficient wireless communication such as the GSM or CDMA networks deployed for mobile phone services, the increasingly popular 802.11x wireless network technologies, Bluetooth and IrDA, 802.15.4, etc. These various networks have intrinsic properties that influence their way of use. QoS is a representing example since according to one approach multiple different levels of QoS can be supported while under a different one, no QoS is considered. Furthermore different technologies assume different coverage ranges as well as expected throughput, implying that a connection using a particular network provides different resources to the devices and applications at the user's disposal. Switching between networks would require session management, creating networking challenges, while moving between dead spots and covered areas results in online and offline statuses. Ideally, these differences must be transparently handled by the interoperability framework, so that the user is not influenced by them. Considering WLANs, the IEEE 802.11x family constitutes a de facto standard owing to its wide deployment, adoption from all kinds of mobile devices and critical advantages providing over competitive options. Recognizing the need for QoS support the e-version of the protocol enhances the protocol performance so as to be able to support such demands and application requirements.

Another critical requirement of any newly introduced device is the ability to use battery or, in general, power sources difficult or even impossible to renew. Furthermore the lifetime requirements extend from hours for laptops, to several days for mobile phones, PDAs, etc. or even months and years for sensor networks, depending on the nodes' capabilities and application demands. Therefore, it goes without saying that energy conservation is of top priority for any architectural design and in this context efficient power management offering power consumption minimization without compromising network performance and maximization of network lifetime is an absolute necessity. The complexity of the goal is increased in more than two hop communications where traffic overburdened nodes and network partitions phenomena must be foreseen and avoided. The main concern is focused on the wireless part, considering that wired stations usually depend on main power sources. However, optimal cooperation implies that wireless as well as wired network must be power aware in order to guarantee transparent and seamless communication.

The importance of power consumption is also shown by the fact that it affects all communication layers and as such demands collaboration between layers, through cross-layer design techniques. At the application layer, network-wide conditions such as congestion and specific channel conditions must be taken in consideration so as to achieve the best possible trade-off between network performance and minimization of power waste. On the other hand, channel arbitration and optimum route selection contribute to the avoidance of hot-spots creation and network partitioning, as well as to the significant decrease of energy consumption when possible. Physical layer optimum RF and ICs design is mandatory in order to achieve low power operation. Even more, in order for any power-aware protocol of network architectural design to show its effect it is necessary to be supported by robust and low power hardware design. Combining these observations it is obvious that only through cross-layer design involving all communication sub-layers and even application layer cooperation can there be truly comprehensive power conservation on a system-wide scale.

In process control applications, where applications are fully distributed and heterogeneity is a major design factor. Most small scale process control systems in operation today, involve simple three-term field controllers embedded in programmable controllers or microcontrollers, operating autonomously and rarely linked to supervisory systems for effective process control. Most large-scale control systems do use Supervision, Control and Data Acquisition (SCADA) or Distributed Control Systems (DCS), few of which, however support optimization. Attempts to introduce high-level supervision that offers optimization of the overall plant, have generally failed due to the lack of coordination and compatibility between the different control levels [43]. Many different classifications of conventional middleware have been developed in order to support distributed and autonomous functionality of objects as part of a whole system for the IoT.

Remote Procedure Call-Based Systems RPC is the most basic form of middleware. It provides an infrastructure to transform procedure calls into remote procedure calls in a uniform and transparent way. Nowadays, RPC systems are used as a foundation for almost all other kinds of middleware.

Transaction Processing Monitors TP (transaction processing) monitors are the oldest and best-know kind of middleware. They are the most reliable, best tested and most stable technology in the enterprise application integration arena. TP monitors are classified into TP-lite and TP-heavy. TP-lite systems typically provide an RPC interface to databases. TP-heavy is a bigger middleware platform with a wealth of tools and functionality that often matches and even surpasses that of operating systems.

Object Brokers When object-oriented languages took over, many platforms were developed to support the invocation of remote objects, thereby leading to object brokers. These middleware platforms were more advanced in their specification than most RPC systems, but they did not significantly differ from them in terms of implementation. In practice, most of them used RPC as their underlying mechanism to implement remote object calls. The most popular object brokers are those based on CORBA (the Common Object Request Broker Architecture), defined and standardized by the Object Management Group (OMG) [32] in 1998.

Object Monitors When object brokers tried to specify and standardize the functionality of middleware platforms, it soon became apparent that much of this functionality was already available from TP monitors. Object monitors are, for the most part, TP monitors extended with object-oriented interfaces.

Message-Oriented Middleware Due to the fact that synchronous interaction was not always needed, TP monitors were extended with persistent message queuing systems. MOM typically provides transactional access to the queues and a number of primitives for reading and writing to local and remote queues.

Message Brokers They are MOMs that have the capability of transforming and filtering messages as they move through the queues. They can also dynamically select message recipients based on the message content. In terms of basic infrastructure, message brokers are just queuing systems with the only difference of that application logic can be attached to the queues, thereby allowing designers to implement much more sophisticated interactions in an asynchronous way.

For constrained-resources systems, non-conventional middleware architectures target the specific needs and particularities of WSN systems.

Mobile Device Information Profile Mobile Device Information Profile (MIDP) is a limited Java-based middleware introduced by Sun for supporting small device [33]. MIDP depends on its own virtual machine, the K Virtual Machine. This virtual machine does not support RMI and object serialization. Consequently, the systems using the MIDP cannot directly communicate with Java Spaces because of missing RMI and serialization capabilities.

SyD is a middleware for mobile devices that has a modular architecture that makes the application development very systematic and flexible. The architecture supports transactions over mobile data stores and provides a persistent uniform object view, as well as several inter-devices communication advantages such as group transaction with QoS specifications or XML vocabulary.

Jadabs was proposed to solve the mobile environment challenges [34]. Its architecture is a lightweight middleware which can be run on a wide range of small devices such as mobile phones, PDAs, motes, or desktop machines. This proposal uses a service-oriented architecture (SOA). For communication between devices, a peer-to-peer infrastructure is used. This enables a communication in decentralized as well as centralized environments.

IAR PowerPac middleware has been developed by IAR Systems [35]. This middleware is planned for constrained-resources embedded devices with little memory space and low-powered microcontroller. IAR PowerPac has a lot of features that increase its usefulness for implementing real-time applications. Among these features are found preemptive scheduling, 255 priorities with an unlimited number of tasks, semaphores, mailboxes, and software timers.

RTZEN is a project of the Electrical Engineering Department of the California University [36]. This project proposes an open-source middleware that fulfills almost all the CORBA 2.3 specification characteristics. Its architecture is based on a model of plug-in layers, which allows that the ORB components, which are not being used, are separated from the ORB core in order to minimize memory space occupied by application. The carrying recipient has been implemented by means of AspectJ.

MicroQoSCorba is a nonstandardized, reduced CORBA implementation [37]. MicroQoSCorba has been designed to support multiple QoS properties, i.e., fault tolerance, security, and timeliness, in constrained-resources systems. This middleware is suited to a wide range of embedded devices owing to its fine granularity of configuration constraints.

Agent-based middleware is based on the idea of an agent, originated by John McCarthy in the mid-1950s. The original view was about a system that could carry out the details of the appropriate computer operations and could ask for and receive advice, offered in human terms, when it was stuck. An agent is, in that sense, a "soft robot" living and doing his business within the computer's world. A more specific definition of "software agent" [38] is a software entity that functions autonomously and continuously in a particular environment. Each agent has incomplete information of capabilities for solving the problem and, thus, has limited viewpoint, with no global system control, data are decentralized, and computation is asynchronous.

Deliberative agent-based architectures assume that the system has an internal model of the world and a planner in charge of deciding which action should be performed next, using predictions of the outcomes of each action [39]. These predictions are based on the world model of the agent. The basis of this architecture is the deliberation, and so, in literature this kind of agents and architecture are also called Beliefs, Desires and Intentions (BDI) agent and BDI architecture. To achieve this objective, environment data are processed in several horizontal layers with different abstraction level. An interpreter decides the actions to achieve the agent goals. Deliberation is required to determine which steps should be taken to achieve the objective.

Because of the bad results obtained with the BDI paradigm, reactive architectures have been proposed, where an agent is specified in terms of two main components: perception and action. Therefore, agent behaviors are the direct reactions to the agent inputs. Another way to see this kind of agents is as finite state machine, which is a model of computation composed of states, transitions and actions.

- States: represent the condition of the agent at a given time.
- Transitions: correspond to a change of state for the object and they are described by a condition that triggers the transition.
- Actions: descriptions of the activities performed at a given time. Depending on its execution, there are two kinds of actions, namely entry and exit.

Recently, mobile agents are in the forefront of research on WSNs, owing to the opportunity and several advantages. Agents are programs that can migrate or clone around the network while maintaining its state. Therefore, hosts are the containers in which agents are located.

Agilla is a mobile agent middleware for sensor networks [40]. Nodes are deployed without specific application. An Agilla application consists of a group of agents injected into the network, therefore, for instance, instead of worrying about how nodes must coordinate to track an intruder, agents can follow the intruder by repeatedly migrating to the node that best detects it. Agents are special processes written in an assembly-like language. The architecture is composed of a core called Agilla Engine that interprets, executes, sends and receives the agents; a dynamic memory manager to share the agents' code, called Code Manager; a Reaction Manager, which allows reactive architectures agents; a TupleSpace Manager to allow agents communicate with each other and a Context Manager to manage the execution state of agents. Agent state, called "Context" by the Agilla developers consists of a data structure, such as microprocessor registers (SP, stack, PC, etc.).

8.7 Wireless Sensor Network Modeling Parameters

WSN parameters are modeled to facilitate end-to-end system performance evaluation through simulation.

8.7.1 Signal Propagation Model

Distance computation is achieved based on the RF propagation model developed in [41], which takes into account environmental parameters and fixed topology impact on the RSSI. As discussed and concluded in [42], for an RSSI-based triangulation technique to be applied in a fixed topology with good results, the mobile target is considered to be at 0.5 m and the fixed nodes at 0.55 m. The RSSI versus distance diagram, adopted in [42], represents a realistic RSSI characteristic.

8.7.2 Network Topology

The network topology has a high impact on the localization accuracy, as fixed nodes positioning defines the communication links quality with the mobile object and thus the localization success. If the fixed nodes are deployed in a grid, the probability that the mobile target is reachable by at least three fixed nodes is high and thus it can be easily located. On the other hand, in cases of random network topology, good communication quality is less likely over the entire localization area, thus resulting in low mobile target connectivity to the fixed nodes and as such degrading localization performance.

8.7.3 Mobile Object Parameters

The two parameters affecting localization accuracy are the speed and the trajectory of the mobile target. If the speed of the mobile target is low, the three nearest nodes have enough time to apply localization calculations and to predict its position. For low mobile target speed, the three fixed neighboring nodes may need more rounds of predictions for increasing the localization accuracy. On the other hand, if the mobile object moves too fast, the time window during which the mobile target moves in the fixed nodes' communication area may not be long enough to allow message exchange and position calculation.

As far as the second parameter is concerned, namely the mobile target trajectory, its impact is lower relative to speed. However, in combination with the fixed nodes topology, it can make communication very difficult and thus the localization task too hard.

8.7.4 HW Characteristics

RFID tagging at pallet-level has been adopted to improve supply chain management by several food and consumer goods retailers. However, product-level tagging has not been widely adopted. Widespread use of RFID at the item- or product-level has been slow due to both cost and privacy concerns. In 2006, the cost of passive RFID tags fell to about 7.9 cents each when purchased in quantities of one million. These are very simple tags providing only product ID. RFID has been used in a number of practical applications, such as improving supply chain management, tracking household pets, accessing office buildings, and speeding up toll collection on roadways. RFID is used to automatically identify people, objects, and animals using short-range radio technology to communicate digital information between a stationary location (reader) and a movable object (tag).

RFID tags fall into two categories; active tags, which contain an internal power source, and passive tags, which obtain power from the signal of an external reader. Because of their lower price and smaller size, passive tags are more commonly used than active tags for retail purposes. A passive tag consists of a microchip surrounded by a printed antenna and some form of encapsulation, plastic laminates with adhesive that can be attached to a product or a small glass vial for implantation. The tag reader powers and communicates with passive tags. The tag's antenna conducts the process of energy capture and ID transfer. A tag's chip typically holds data to identify an individual product, the product model, and the manufacturer.

In contrast to simple tags used for product identification, complex tags support data logging, remote sensing, and other functionality. These types are much more complex, larger in silicon size, and therefore more expensive (above 1 EUR). Because of superior functionality, these tags consume more power and in case of logging application need battery support.

Although RFID applications have become more widespread in recent years, radio frequency identification is not a new technology. Although prices have fallen, placing RFID tags on individual items may still be impractical for many inexpensive consumer products, but could be cost effective for more expensive items like clothing. Existing RFID technology has the following limitations regarding goods tagging that need to be addressed:

- Lack of communication stability in harsh environment
- Insufficient communication distance
- Too large power consumption of the active tags, mainly for supplying the nonvolatile on-board memory

8.7.5 Sensing and Energy Conversion

The state of the art in advanced goods tracking is based on individual item's tracking throughout the production chain using UHF RFID technology that is electronic product code (EPC) Class 1 Generation 2 compatible. No other tracing of condition variables, although extremely essential for quality of goods, were monitored as much as moisture and temperature. In the case of food tracing, an individual meat package can be traced back to a day batch of a slaughterhouse. The field would benefit from the development of systems enabling, for example, farm-level traceability of meat. Specific farms may have certificates, etc., related to producing specific products.

One of the most important individual aspects followed and monitored throughout the production chain is temperature. The food industry is using temperature loggers to follow the cold chain during manufacture and transportation to customers. Improvement in following temperatures during various phases is of interest.

Among others, the logger data is not always conveniently readable for the manufacturer at the customer end of the chain. In addition to cold chain, temperature logging in various cooking and pasteurization phases is not always convenient, for example, some microwave devices do not enable the use of temperature loggers.

An exploratory study was made in 2008, where two nails of different metals were inserted into wood and connected electrically. The setup comprised of an electrochemical cell where the nails acted as electrodes and wood as electrolyte. Acceptable currents (0.5–3.5 μA) were obtained as long as the wood moisture was kept relatively high. The battery voltage was low, 0.3–0.8 V, and depended on factors as the insertion depth of the nails and the distance between the electrodes.

8.7.6 Process and Supply Chain Modeling and Evaluation

Innovative solutions and innovations developing the next generation of transponder systems will increase the ability to collect more and more data. This implies new ways of modeling techniques to transform data into information. These modeling techniques will also be developed to interface with the transponders on one hand and the HMI on the other. Empirical, multivariate statistical, models have gained more and more use in all types of industries for powerful evaluation of Mega Variate data during the last 20 years. This can be attributed to: more computing power which enables more models to be tested and validated in search of a suitable model, the access to historical data through efficient specialized databases, more intuitive software, as well as more widely spread knowledge about these methods.

The classical role of evaluation of process data being generated along the production chain are typically based on set values, upper and lower limits, etc. Multivariate Statistics Process Control (MSPC) techniques such as principal component analysis (PCA) has proven useful for both fault detection and isolation [44–48] and can be considered an important tool to aid in the interpretation and decision making process of the plant engineer. PCA reduces the dimensionality of data, which can be crucial for handling large data quantities and also offers a method to categorize process and equipment deviations.

Partial Least Squares (PLS) [49] can be used for multivariate calibration and supervised classification of data using a reference variable or several that may be either continuous or binary.

This could be helpful when finding undiscovered patterns. Apart from being a tool for process monitoring, multivariate analysis can be used to create soft-sensors, replacing difficult or expensive online measurements and offering a possibility to validate the performance of existing sensors.

No industrial or scientific publications using active transponders with specific food industry sensors in combination with multivariate analysis have been found in the literature. Models would enable more accurate and realistic estimation of raw material quality and product shelf life during the whole distribution chain. Special treatments are needed to use categorical data such as quality classes or other binary representations in multivariate models such as PCA or PLS23.

References

1. Gautier, P. (2007). RFID et acquisition de données évènementielles: retours d'expérience chez Bénédicta, pages 94 à 96, Systèmes d'Information et Management—revue trimestrielle N°2 Vol. 12, ISSN 1260-4984/ISBN 978-2-7472-1290-8, ESKA.
2. Guo, B., Zhang, D., & Wang, Z. (2011). Living with IoT: The emergence of embedded intelligence. International conference on and 4th International Conference on Cyber, Physical and Social Computing.
3. Waldner, J.-B. (2007). *Nanoinformatique et intelligence ambiante. Inventer l'Ordinateur du XXIeme Siècle* (p. 254). London: Hermes Science. (ISBN 2-7462-1516-0).
4. Open Geospatial Consortium. OGC Abstract Specification.
5. Paul, S., Evans, J., & Raychaudhuri, D. (2006, September). Technical document on overview wireless mobile and sensor networks. GDD-06-14-GENI: Global Environment for Network Innovations.
6. Lo, B., et al. (2005). Body sensor network—A wireless sensor platform for pervasive healthcare monitoring. Adjunct Proceedings of the 3rd International Conference on Pervasive Computing (PERVASIVE 2005), pp. 77–80.

7. The Project HEARTS (Health Early Alarm Recognition and Telemonitoring System). http://heartsproject.datamat.it/hearts/Sections/01The%20Project.
8. Van Laerhoven, K., Lo, B. P. L., Ng, J. W. P., Thiemjarus, S., King, R., Kwan, S., Gellersen, H., Sloman, M., Wells, O., Needham, P., Peters, N., Darzi, A., Toumazou, C., & Yang, G. (2004). Medical healthcare monitoring with wearable and implantable sensors. UbiHealth, 6–7th of September 2004.
9. Shea, D. A., & Lister, S. A. (2003, November). The biowatch program: Detection of bioterrorism. Congressional Research Service Report RL 32152, Science and Technology Policy Resources, Science and Industry Division.
10. Anderson, E., Girard, T., & Ottavianelli, G. (2003). A micro-satellite and in situ ground sensor network for combating malaria. In Proceedings of 54th International Astronautical Congress.
11. Smart Medical Home at the University of Rochester. http://www.futurehealth.rochester.edu/smart_home/.
12. Sanders, J. (2000). Sensing the subtleties of everyday life. Research Horizons Magazine 17, no. 2, Georgia Institute of Technology.
13. Defense Advanced Research Projects Agency (DARPA). (1997). Joint Program Steering Group Arlington, Virginia. NISE east electronic security systems engineering division North Charleston, South California. Perimeter Security Sensor Technologies Handbook (online manual). http://www.nlectc.org/perimetr/start.htm.
14. Manufacturer of Barricade-500. Advanced outdoor vibration detection system. http://www.magal-ssl.com/products/?pid=19.
15. Zhao, F., & Guibas, L. (2004). *Wireless sensor networks: Information processing approach*. Elsevier.
16. Coleri, S., Yiu, S., & Varaiya, P. (2004, September). Sensor networks for monitoring traffic. Invited paper Allerton Conference.
17. Conant, R. (2006, spring). Wireless sensor networks: Driving the new industrial revolution. Industrial Embedded Systems.
18. Low, K. S., Win, W. N. N., & Er, M. J. (2005). Wireless sensor network for industrial environments. Proceedings of the 2005 International Conference on Computational Intelligence, Control and Automation.
19. Belén, A., García, H., Martínez-Ortega, J.-F., López-Navarro, J.-M., Prayati, A., Redondo-López, L. (2008, November). *Problem solving for wireless sensor networks*. London: Springer. (ISBN: 978-1-84800-202-9).
20. Eddie, B. S., Tan, J. G., Winston, L., Seah, K. G., & Rao, S. V. (2006). On the practical issues in hop count localization of sensors in a multihop network. Proceedings of the 63 rd IEEE Vehicular Technology Conference (VTC2006-Spring), 8–10 May.
21. Chen, Y.-S., Lee, S. L. G., Lin, T.-H. (2005, April). PCAR: A power-aware chessboard-based adaptive routing protocol for wireless sensor networks. Journal of Internet Technology, Special Issue on Wireless Ad Hoc Network and Sensor Networks.
22. Zhou, J., & De Roure, D. (2006, June). Designing energy-aware adaptive routing for wireless sensor networks. Proceedings of International Conference on ITS Telecommunications, pp. 680–685.
23. Heinzelman, W., Chandrakasan, A., & Balakrishnan, H. (2000, January). Energy-efficient communication protocols for wireless microsensor networks. In Proceedings of Hawaaian International Conference on Systems Science.
24. Heinzelman, W., Kulik, J., & Balakrishnan, H. (1999, August). Adaptive protocols for information dissemination in wireless sensor networks. Proceedings of 5th ACM/IEEE Mobicom Conference, Seattle, WA.
25. Polastre, J., Hill, J., & Culler, D. (2004). Versatile low power media access for wireless sensor networks. SenSys '04, ACM, November 3–5, Baltimore Maryland, USA.
26. Suh, C., Ko, Y.-B., & Son, D.-M. (2006). *An energy efficient cross-layer MAC protocol for wireless sensor networks* (APWeb 2006, LNCS 3842, pp. 410–419). Berlin: Springer.
27. Warrier, A., Min, J., & Rh, I. Z-MAC: a Hybrid MAC for wireless sensor networks. SIGCOMM '05 ACM, USA.

28. Madan, R., Cui, S., Lall, S., & Goldsmith, A. (2005). Cross-layer design for lifetime maximization in interference-limited wireless sensor networks. In Proceedings of IEEE INFOCOM '05, vol.3, pp. 1964–1975.
29. Dong-Mei, S., & Bing, H. E. (2006, November). Review of key management mechanisms in wireless sensor networks. *Acta Automatica Sinica, 32*(6).
30. Camptepe, S. A., & Yener, B. (23 March 2005). Key distribution mechanisms for wireless sensor networks: A survey. Technical Report TR-05–07.
31. Potlapally, N. R., Ravi, S., Ranghunathan, A., & Jha, N. K. (2006, February). A study of the energy consumption characteristics of cryptographic algorithms and security protocols. *IEEE Transaction on Mobile Computing, 5*(2).
32. OMG. (1998, February). The common object request broker: Architecture and specification, Object Management Group.
33. Sun Microsystems. (2007). http://java.sun.com/products/midp/.
34. Frei, A. R. (2005). Jadabs: An adaptive pervasive middleware architecture.
35. Systems, I. A. R. (2007). http://www.iar.com/.
36. Panahi, M., et al. (2007). http://zen.ece.uci.edu/rtzen/.
37. McKinnon, A. D., et al. (2002). MicroQoSCORBA: A QoS-enabled, reflective, and configurable middleware framework for for embedded systems. OMG Real-time Workshop.
38. Shoham, Y. An overview of agent-oriented programming. In J. M. Bradshaw (Ed.), *Software Agents*. Menlo Park: AAAI Press.
39. Bursie, C. I. *Dog-like behaviour selection for an AIBO robot dog*. Stockholm, Sweden: Royal Institute of Technology.
40. Fok, C., Roman, G., & Lu, C. (2006). *Agilla: A mobile agent middleware for sensor networks*. St. Louis: School of Engineering and Applied Science. Washington University.
41. Stoyanova, T., Kerasiotis, F., Prayati, A., & Papadopoulos, G. (2007). Evaluation of impact factors on accuracy for localization and tracking applications. 5th ACM international workshop, MOBIWAC.
42. Kerasiotis, F., Stoyanova, T., Prayati, A., & Papadopoulos, G. (2008). A topology-oriented solution providing accuracy for outdoors RSS-based tracking in WSNs. Proceeding of SENSORCOMM '08. Proceedings of the 2008 Second International Conference on Sensor Technologies and Applications.
43. Samad, T. (1996, December). Intelligent control in the process industries: considerations for future research. Proceedings of the 35th Conference on Decision and Control, Kobe, Japan, pp. 4512–4513.
44. Dunia, R., & Qin, S. J. (1998). Joint diagnosis of process and sensor faults using principal component analysis. *Control Engineering Practice, 6*(4), 457–469.
45. Albazzaz, H., Wang, X. Z., & Marhoon, F. (2005). Multidimensional visualisation for process historical data analysis: a comparative study with multivariate statistical process control. *Journal of Process Control, 15*(3), 285–294.
46. Lee, C., Choi, S. W., & Lee, I. (2004). Sensor fault identification based on time-lagged PCA in dynamic processes. *Chemometrics and Intelligent Laboratory Systems, 70*(2), 165–178.
47. Villez, K., Steppe, K., & De Pauw, D. J. (2009). Use of Unfold PCA for on-line plant stress monitoring and sensor failure detection. *Biosystems Engineering, 103*(1), 23–34.
48. Wise, B. M., & Gallagher, N. B. (1996). The process chemometrics approach to process monitoring and fault detection. *J Proc Control, 6*(6), 329–348.
49. Geladi, P., & Kowalski, B. R. (1986). Partial least-squares regression: A tutorial. *Analytica Chimica Acta, 185,* 1–17.

Chapter 9
Shape Analysis in Radiotherapy and Tumor Surgical Planning Using Segmentation Techniques

S. Zimeras, L. Gortzis and Ch. Pylarinou

Abstract Most medical imaging applications base their functionality on replacing the physical patient model with the digital data of the patient coming from any medical imaging modality. Medical imaging techniques offer unique capabilities on collecting digital data of the human body. Nowadays, technical evolutions allow the generation of three-dimensional (3D) data within a few moments. The 3D dataset has a great benefit over conventional two-dimensional (2D) images, especially in cases with complex anatomy or pathology. Radiotherapy treatment (RT) is a very demanding cancer treatment process. The aim of the treatment is to cure or to limit the disease with a minimum possible damage of healthy tissues. The process is composed of several steps that are highly dependent on each other in order to achieve the desired results. Quantitative analysis of digital images requires detection and segmentation of the borders of the object of interest. Accurate segmentation is required for volume determination, 3D rendering, radiation therapy, and surgery planning. In medical images, segmentation has traditionally been done by human experts. Substantial computational and storage requirements become especially acute when object orientation and scale have to be considered. Therefore, automated or semiautomated segmentation techniques are essential if these software applications are ever to gain widespread clinical use. Many methods have been proposed to detect and segment 2D shapes, most of which involve template matching. Advanced segmentation techniques called Snakes or active contours have been used, considering deformable models or templates. The main purpose of this work is to apply segmentation techniques for the definition of 3D organs (anatomical structures) under RT.

9.1 Introduction

Radiotherapy treatment (RT) is the most important technique to identify the shape of the tumor in the organ that the doctors are interested in. This technique is widely used for cancer treatment. A major consideration for RT application is the volume

S. Zimeras (✉)
University of the Aegean, Karlovassi, 83200 Samos, Greece
e-mail: zimste@aegean.gr

L. Gortzis · Ch. Pylarinou
University of Patras, 26504 Rio, Greece

N. Sklavos et al. (eds.), *System-Level Design Methodologies for Telecommunication*, 159
DOI 10.1007/978-3-319-00663-5_9, © Springer International Publishing Switzerland 2014

visualization of medical data, where a three-dimensional (3D) shape model is reconstructed by combining all the two-dimensional (2D) images of cross sections of the organ. On the basis of that shape, doctors can identify the position, shape, and importance of tumor, especially when a radiotherapy dose must be applied. The 3D visualization of these organs most of the times requires effective segmentation techniques for the identification of the shapes [35].

Classically, image segmentation denotes the technique of regional area identification based on various splitting method extaction structures representing, as physically as possible, the real image data (2D regions or 3D voxels). Image segmentation may be performed in one of these ways: manual segmentation methods where physical hand (like doctor) reconstructs the segmentation regions; fully automatic segmentation methods where effective segmentation algorithms without any other involvement identify the regions of interest; and semiautomatic segmentation methods, which combine both manual and automatic methods, where a starting point is most of the times enough for the marking of the segment regions. Methods such as boundary tracking (BT), region growing, edge detection, active contours, and deformable models [32] are part [16] of the most important segmentation techniques that could be used (as it is or combined together) to perform a shape segmentation process.

On the basis of any segmentation technique, contours can adapt the shape of the organ, considering the boundaries of the anatomic structures of interest. Contours are defined on the basis of points. For the better representation of shape of the organ (2D or 3D presentation), application of interpolation techniques is necessary. The connectivity between points can be linear or higher order. The higher-order connectivity can be achieved using interpolation models such as Hermite cubic splines, spline curves, and Bezier curves [8, 20], which are the most common and successfully used techniques for smoothing curves' RT planning. The aim of the high-order interpolation techniques is to reduce the amount of input points required to describe a smooth shape. By performing a simple comparison between linear and higher-order interpolation, we can find out that the amount of points used to illustrate the shape of a structure. Linear interpolation requires at least twice as many samples as higher-order interpolation algorithms. A common methodology used to combine high-order interpolation and image edge properties is the use of active contour models. The active contour models or Snakes can be 2D image curves [6, 17] or 3D polygon meshes [31], which are adjusted from an initial approximation to the image or volume features by a movement caused by simulated forces. Image features provide the so-called external force. An internal tension of the curve resists against highly angled curvatures, and this makes the Snake movement robust against noise. After a starting position is given, the Snake adapts itself to shape by relaxation to the equilibrium of the external force and internal tension. Snakes have been proven efficient and fast for a number of applications in medicine involving different imaging modalities [4, 10, 11, 22, 29]. Snake models are a class of energy-minimizing spline curves or surfaces. These models are very important in a number of inverse visual problems such as the segmentation and reconstruction of objects from images in mathematical ill-posed problems.

The output result of the contour reconstruction algorithm is provided in two forms: as a triangulated mesh or as multiple parallel contours extracted in arbitrary plane directions. For the RT planning applications, we focus mostly on the reconstruction of contours in the axial direction. The algorithm can be separated into different modules from the point editing to the reconstruction of the surface and contours as follows:

1. Collection and processing of the given contour points generating contour.
2. Calculation of the implicit functions in 3D by solving a linear equation system that will give us the coefficients of the interpolation function.
3. Evaluating the implicit function over a 2D planar contours or 3D polygon meshes.

9.2 Radiotherapy Treatment Planning

The goal of RT planning is to find an effective way to deliver an adequate irradiation dose to the target volume without causing severe damage to surrounding normal tissues. An interactive optimization of treatment plans could be achieved by combining target volume, organ at risk, and simulated radiation dose in a 3D visualization system [9, 18, 30], which most of the times is a 3D virtual simulator [13]. On the basis of virtual simulation system, the procedure for radiotherapy planning is as follows:

1. The patient is moved to the computed tomography (CT) scanner.
2. Image data (3D shape of the organ) and target region are transferred to the virtual simulator.
3. The doctor starts the real treatment of the patient on the basis of real-time virtual simulator.

Considering the virtual simulator, the planning process can be divided into following steps:

1. Determination of the target volume and organs at risk
2. Virtual therapy simulation
3. Dose calculation
4. Visualization and evaluation of dose distributions

The goal of treatment planning is the determination of a suitable and practicable irradiation technique, which results in a conformal dose distribution; thus, treatment planning is a typical optimization problem. In addition to static treatment planning, volumetric image data is playing a larger role in treatment delivery and adaptive radiotherapy. For that reason, volume data must be defined as important part for the 3D voxel reconstruction of the shape of the organs. A volume dataset is a set of discrete sample points in an object space, $V(x)$, $x \in R^n$, in which $\{x\}$ is a set of sampling points; n is the dimension of the sampling space (usually $n = 3$, i.e., 3D volume data); and V represents the sampling values. In medical imaging, volume data is usually a structured dataset, typically organized as a stack of slices; V can be single- or multivalued. The more traditional surface-extraction methods first create

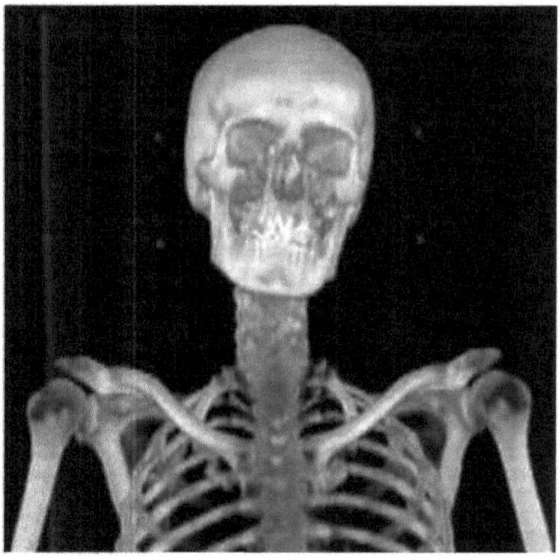

Fig. 9.1 Specific gray scale tissue image

an intermediate surface representation of the objects on the basis of the volume rendering [21, 28]. Volume rendering techniques use attenuation, color, scattering, and movement to generate images, where the Hounsfield values could be used to identify the brightness of every voxel given by

$$HU = \left(\frac{HU_{\max} - HU_{\min}}{We - Ws}\right)(Wl - Ws) + HU_{\min},$$

where $[HU_{\max}, HU_{\min}]$ is the gray-level range, $[Ws, We]$ defines the window width, and Wl defines the window center. Figure 9.1 represents a resulting image based on displaying specific tissue type in the full range of the gray scale, applying the Hounsfierld window technique.

To overcome the problem with the shading of the images, where anomalies must be reproduced, a Z-buffer gradient estimation technique is applied, without using the 3D data. A depth map can be illustrated as a function of two variables $z = (x, y)$ and the surface could normally be obtained by the gradient vector ∇z. The partial derivates $\frac{\partial z}{\partial x}$ and $\frac{\partial z}{\partial y}$ define the image $n = \left(-\frac{\partial z}{\partial x}, -\frac{\partial z}{\partial y}, 1\right)$.The distance z_{ij} can be calculated by the derivates, where i and j are indices specifying the location of a pixel in the Z-buffer. The derivate $\frac{\partial z}{\partial x}$ may be calculated by the forward difference $d_f = z_{i+1,j} - z_{i,j}$ or by the backward differences $d_b = z_{i,j} - z_{i-1,j}$ or by the average of d_f and d_b, $d_{\text{aver}} = \frac{1}{2}(z_{i,+1,j} - z_{i-1,j})$,which is defined as the central difference. The central difference is the better approximation if z is a continuous function of x. Figure 9.2 illustrates the example where target organ and projection of treatment field using Z-buffer methodology are applied [28].

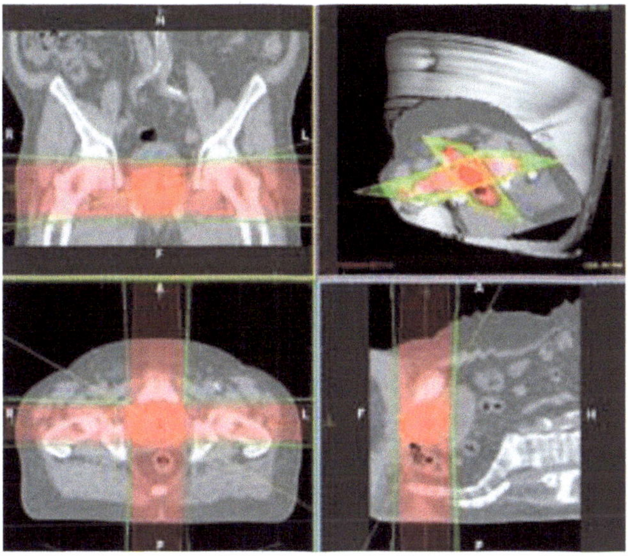

Fig. 9.2 3D project treatment field with patient 3D data using Z-buffer technique. *Red* object (contour): tumor

9.3 Segmentation Techniques

Segmentation is the process where the image is divided in regions based on particular color indices and the shape of the particular organ is represented in a better way. This converts the planar pixel of the image into a distinguishable number of individual organs or tumor that can be clearly identified and manipulated. The segmentation process may involve complicated structures, and in this case, usually only an expert can perform the task of manual identification on a slice-by-slice basis. Humans can perform this task using complex analysis of shape, intensity, position, texture, and proximity to surrounding structures. All these features are differently qualified depending on the experience of the user. To generate a "complete" segmentation application, numerous tools and algorithms must be combined [19]. We will investigate the volume segmentation aspects within the frame of radiation therapy.

The mathematical formulation of segmentation is defined as follows [27]: Let I be the set of all image pixels, then by applying segmentation, we obtain different unique nonoverlapping regions $\{S_1, S_2, \ldots, S_n\}$, which, when combined, form I: $\bigcup_{i=1,n}^{n} S_i = I$, where $S_i \cap S_j = O S_i$ is a connected region $i = 1, 2, \ldots, n$, $P(S_i) = \text{TRUE}$ for $i = 1, 2, \ldots, n$, $P(S_i \cup S_j) = \text{FALSE}$ for $i \neq j$, and $P(S_i)$ is a logical predicate defined over points in set S_i.

The easiest segmentation technique is the manual consideration where the doctor reconstructs and manipulates the contours of the specific organ that he/she is interested in. This technique is time consuming and the doctor must be very experienced

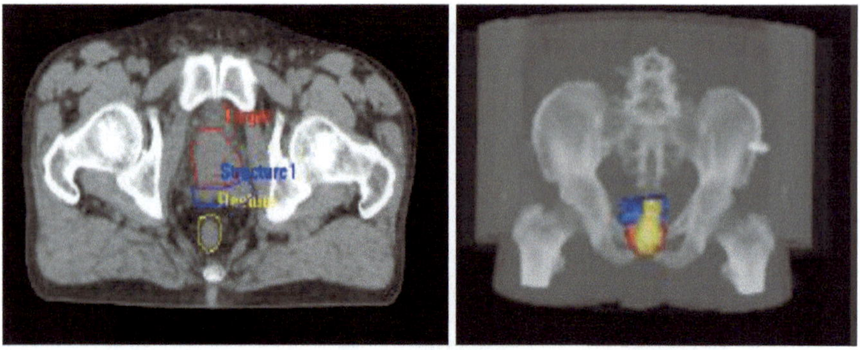

Fig. 9.3 2D regions and 3D volume data for prostate cancer using manual segmentation

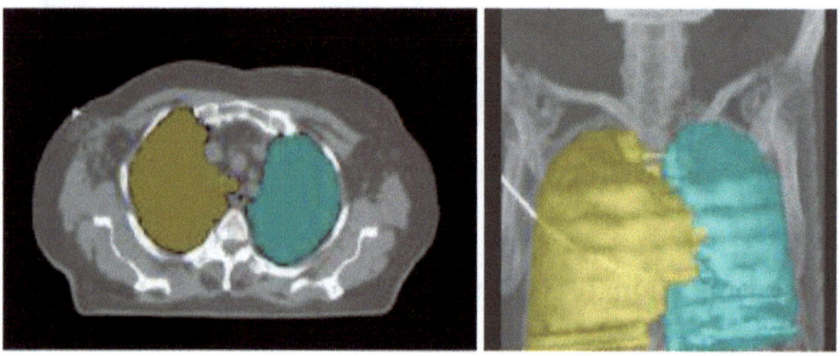

Fig. 9.4 Region growing results for the left and right lungs [15, 16]

to draw the contours as well as to manage the particular electronic system. Figure 9.3 illustrate an example of manual segmentation for the prostate cancer for 2D regions and 3D volume data.

Another effective technique is the region growing [1], which is based on thresholds, most of the times based on the gray scales of the image given by the Hounsfield values. Particularly, the method starts with a group of voxels that are reconstructed in the organ defining the starting seed. The seed then grows in a specific way, joining neighboring voxels that are similar. Adams and Bischof [1] propose a method that includes all the image voxels for different seeds, giving as many regions as the number of seeds. The method is effective with very good results but the drawback is when the edges of the image are not clearly recognized or the organ that we are interested in has a complicate convex shape. Figure 9.4 illustrates region growing results (2D regions and 3D volume) for the left and right lungs.

BT technique [14, 15, 36, 37] is a semiautomatic segmentation method that given one point along a region's [16] boundary follows the boundary around the region until it returns to the original point. The process starts with a starting point for a specific direction. When a sharp edge is found, the specific point is marked and the process

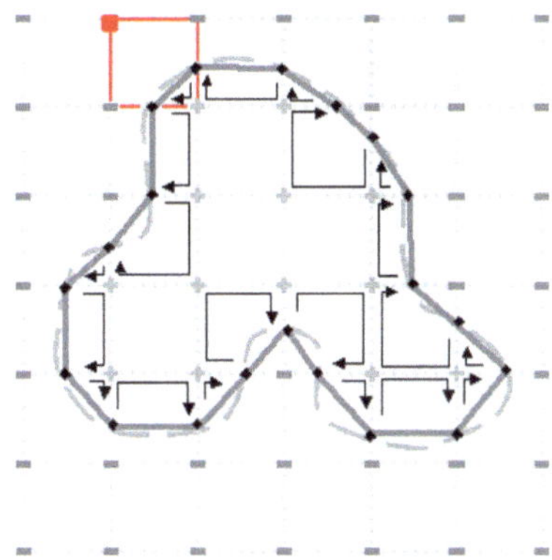

Fig. 9.5 Boundary tracking. The *red point* is the starting position

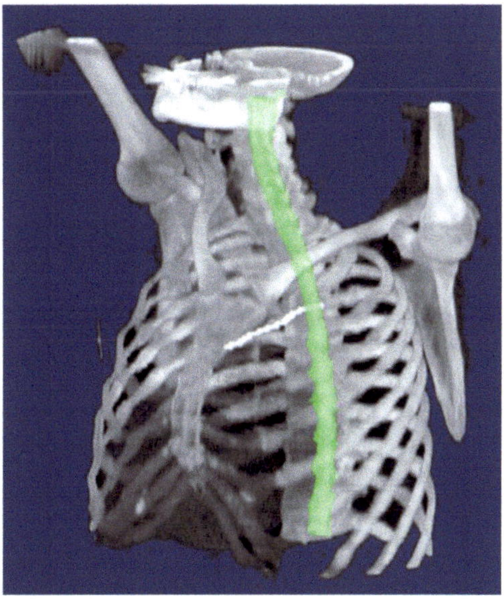

Fig. 9.6 Boundary tracking method for the spinal canal

continues in a different direction. When all the points that represent the appropriate shape have been marked, the algorithm stops (Fig. 9.5).

The advantage of the approach is that no assumptions are needed to be made a priori about the boundary shape that may vary from a straight line to much more complex shapes, which are difficult to parameterize. Figure 9.6 illustrates the results for the BT method for the spinal canal organ.

Major problem with the spinal canal was the shape of the organ. The original algorithm did not give us satisfactory results with inaccurate segmentation contour reconstruction (Fig. 9.7).

For that reason, a radial function was used with the main purpose to find the sharp edge of the organ. Radial function $R(\omega)$ is a contour-based polar shape representation,

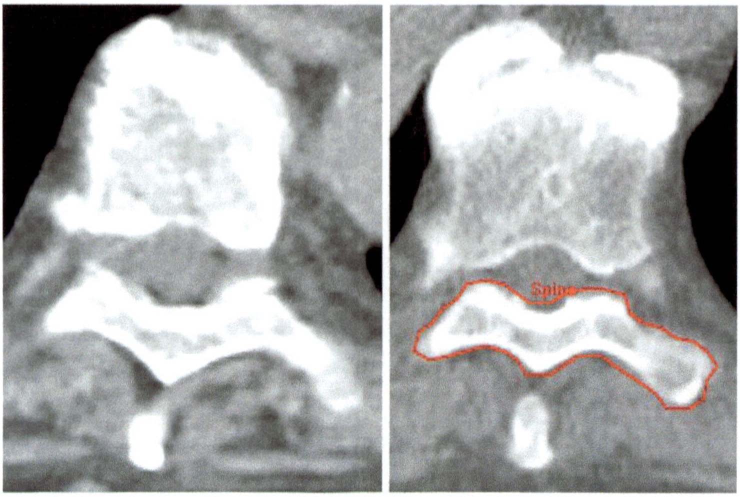

Fig. 9.7 Original image and inaccurate segmentation for spinal canal using boundary tracking method

Fig. 9.8 Radial function

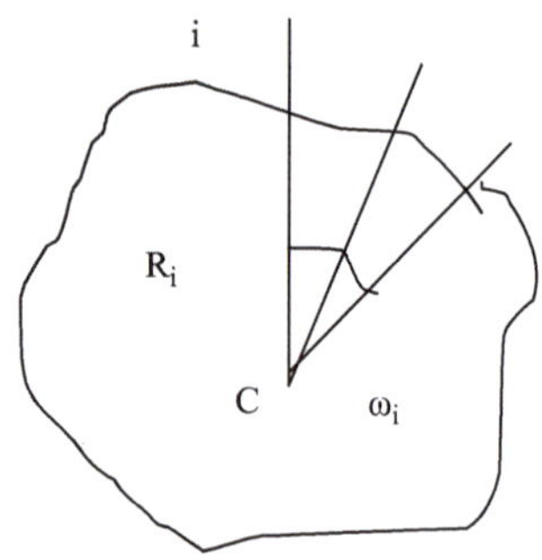

which shows the distance R between an interior point (centroid C) and the contour points as a function of the polar angle ω (Fig. 9.8).

So far, only a very few techniques propose to the physicians one or more alternatives for volume definition [5]. Recently, Pekar et al. [26] reported a method based on an adaptation of 3D deformable surface models to the boundaries of the anatomic structures of interest. The adaptation was based on a trade-off between deformations of the model induced by its attraction to certain image features and the shape integrity of the model. Nevertheless, to make the concept clinically feasible, interactive tools were also introduced that allow quick correction in problematic areas in which the automated model adaptation may fail. Active contours or Snakes are part of the deformable models. The idea behind these models is to represent the contours as parts of elasticity and rigidity. The general concept of active contours is autonomous adaptation of the shape and location of objects finding the important contour points to reconstruct the image. After initialization, by a starting contour, the process performs a fitting process based on the elasticity of the contour lines. If the model is represented by a parametric curve in 2D, then $v(s) = (x(s), y(s))$. The deformation is performed by minimizing an energy function E_{snake} that integrates image features, shape, and internal constraints.

$$E_{\text{snake}} = \int_0^1 E_{\text{int}}(v(s)) + E_{\text{image}}(v(s)) + E_{\text{con}}(v(s))ds.$$

Terzopoulos [31, 32], assuming that a contour could be identified by point $P_i = (x_{i,}\ y_j)$, $i = 0, 1, 2, \ldots, n-1$, introduces the elasticity with energy forcing the curve to shrink to a single point by

$$E_{\text{int}-\text{elasticity}} = a\sum_{i=o}^{n-1} L_i^2 = a\sum_{i=o}^{n-1} (x_{i+1} - x_i)^2 + (y_{i+1} - y_i)^2.$$

The image stiffness energy (for the smoothness after the minimization) with

$$\begin{aligned} E_{\text{in}-\text{stiffiness}} &= \beta\sum_{i=o}^{n-1} |P_{i+1} - 2P_i + P_{i-1}|^2 \\ &= \beta\sum_{i=1}^{n-2} (x_{i+1} - 2x_i + x_{i-1})^2 + (y_{i+1} - 2y_i + y_{i-1})^2, \end{aligned}$$

with gradient energy given by

$$E_{\text{ext}-\text{grad}} = -\lambda_1 \sum_{i=0}^{n-1} \left(\left|G_x(x_i, y_j)\right|^2 + \left|G_y(x_i, y_j)\right|^2\right).$$

The internal energy describing how well the curve is fitted to the image data is given by

$$E_{\text{internal}} = aE_{\text{in}-\text{elasticity}} + \beta E_{\text{in}-\text{stiffiness}}.$$

Finally, the energy for the constraints is given by

$$E_{\text{constraints}} = \eta \sum_{i=0}^{N-1} \min |P(t) - P|,$$

where P are the point constraints.

Results for the process are given in Fig. 9.9.

Grosskopf et al. [9, 10] proposed a model based on Lagrangian dynamics (LD) approach. The curve is moved by the application of two forces: acceleration force F

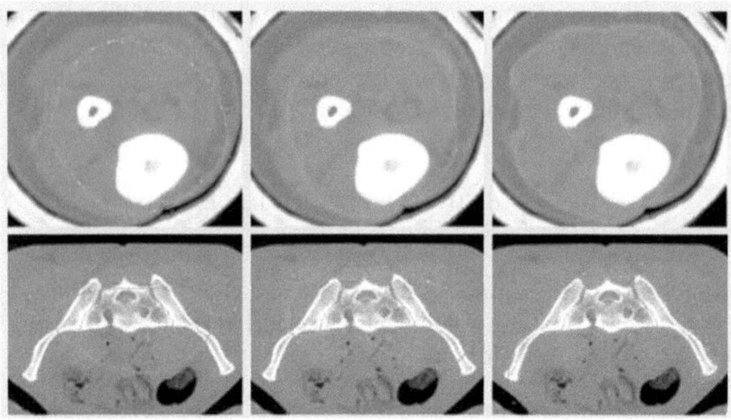

Fig. 9.9 Active contour tracking

and damping force D. The acceleration, velocity, and curve position are calculated iteratively using an explicit time integration scheme given by

$$\frac{\partial^2 v(s)}{\partial^2 t} = \frac{F(v(s))}{m(s)}$$
$$\frac{\partial v(s)}{\partial t} = \frac{\partial v(s)}{\partial t} + \Delta T \frac{\partial^2 v(s)}{\partial^2 t} - D(v(s))\frac{\partial v(s)}{\partial t}$$
$$v(s) = v(s) + \Delta T \frac{\partial V(s)}{\partial t},$$

where ΔT is the size of time interval (Fig. 9.10).

The interpolation techniques described earlier are usually applied only on a single-slice level (2D). The use of high-resolution CT data allows the use of multiplanar reconstructions (MPR) for the sagittal and coronal direction, in relation with the patient anatomy. These two images are orthogonal to each other and perpendicular with the axial plane. The navigation through these images helps in the observation of complex anatomy. The sagittal and coronal views often offer a better overview of organs' 3D shape. Defining volumes in these directions could be of benefit, since several organs are aligned along the longitudinal body axis. The problem we have to solve in our case is the generation of a surface and contours from structured closed parallel and nonparallel contours. The data points represent the contour points as they are generated by the user on the different levels of the axial slices. The most common approaches used to reconstruct surfaces form parallel contours are well established in medical imaging applications [3, 7, 23, 25, 34]. The limitation of these algorithms is that they cannot be applied on nonparallel contours. The problem of the nonparallel contours could also be formulated as generation of surfaces from scatter data, which are very common in industrial applications [2, 12]. For our application, we selected the approach presented by Turk and O'Brien [33]. Their

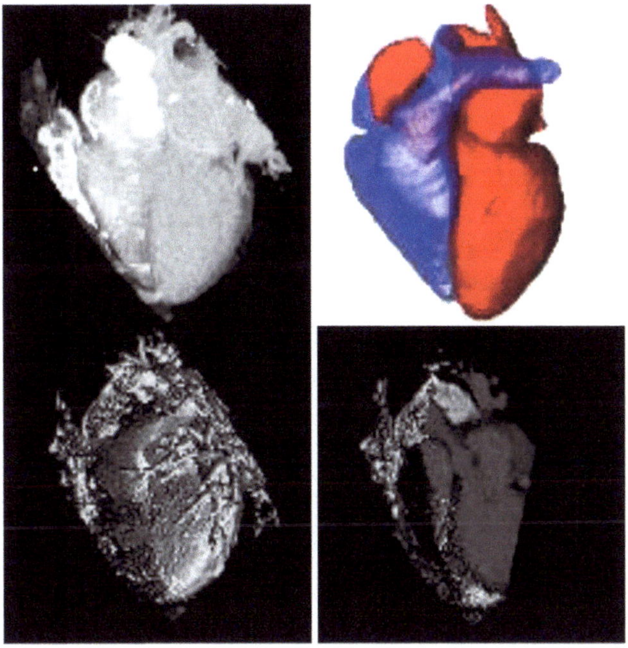

Fig. 9.10 Reconstruction of the heart based on the active contours models [10, 11]

method adapts earlier work on thin-plate splines and radial basis interpolants to create a new technique in generating implicit surfaces.

Their method allows direct specification of a complex surface from sparse, irregular, scatter samples. The main restriction of the method is the relatively small number of sample data that can be handled. This drawback makes the above approach unsuitable for a number of applications, as a large number of sampling points are needed. At this point, we demonstrate the use of implicit function interpolation to reconstruct 3D organ shapes that have been defined selectively by the user. The total number of contour points will be used as the input data to the implicit surface algorithm with arbitrary order. The number of these sampling points will not exceed the level of few hundred, and therefore the calculation times will be in acceptable ranges despite the complexity of the algorithm.

Voronoi diagrams could be applied to construct the shape of the organ with a lot of anomalies. Let P be a set of n points $p_0, p_1, \ldots, p_n$ in R^2, with $n \geq 3$ and with no more than three cocircular points. The Voronoi diagram Var(P), also known as the Dirichlet tessellation, is the partition of the plane in n convex regions, such that each region contains one point $p_i \in P$ and all the points in R^2 that are closer to p_i than to any other point $p_j \in P$ [24].

A Voronoi region is defined as $VR_i = \{q, q \in R^2 | d(q, p_i) < d(q, p_j); p_i \neq p_j \in P\}$, where $d(q,\ p_j)$ is the Euclidean distance. A Voronoi edge is related to Voronoi regions with $VE_{ij} = \{q, q \in R^2 | d(q, p_i) = d(q, p_j); p_i, p_j \in P; \nabla p_k \in P,$

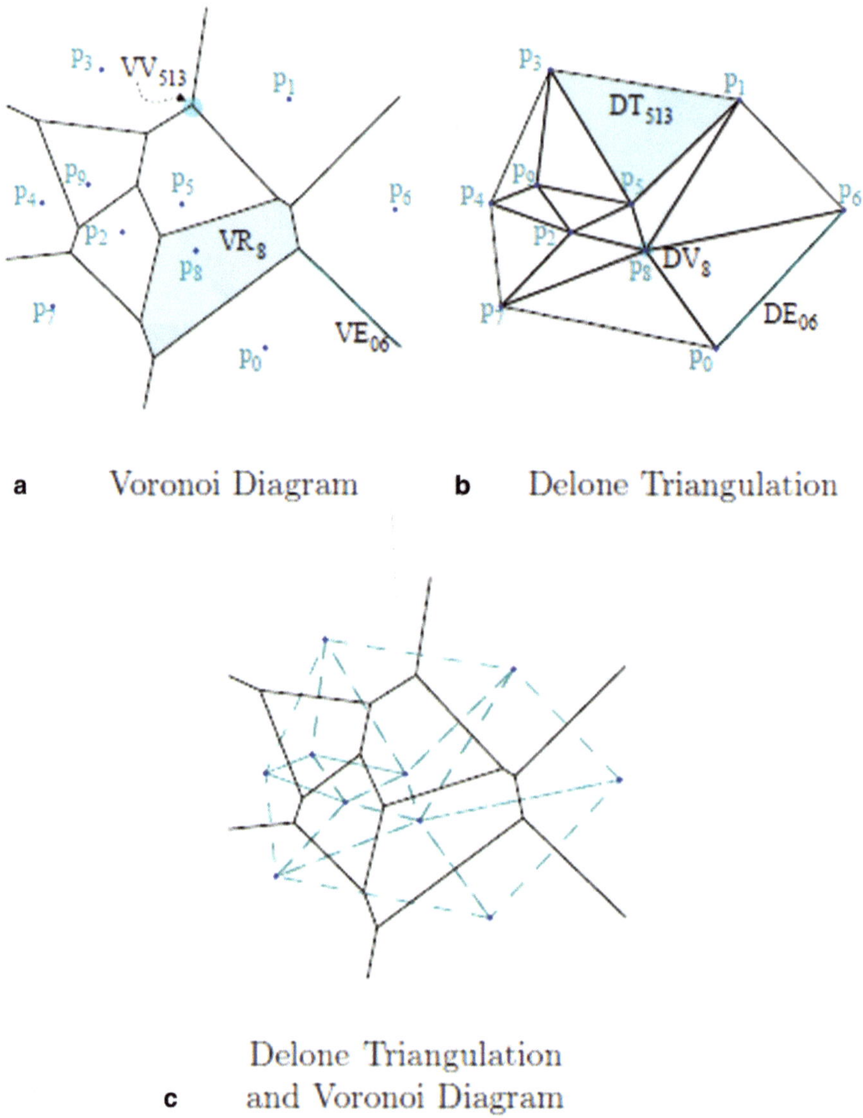

Fig. 9.11 The 2D Voronoi diagram and the 2D Delone triangulation of 10 points in R^2 [24]

$p_k \neq p_j, d(q, p_i) < d(q, p_k)\}$. For a 3D reconstruction of a shape based on the contours, a combination between regions and edges must be achieved where a Delone triangulation could be implemented (Fig. 9.11). Results of combination of all the slices of the organ that we are interested in are given in Fig. 9.12.

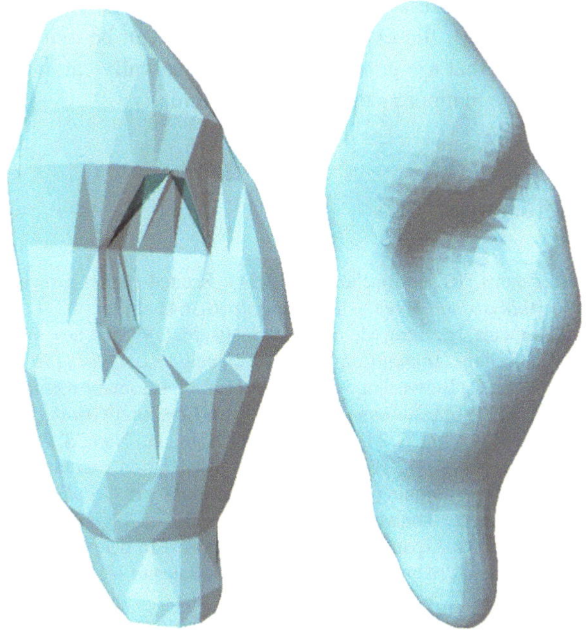

Fig. 9.12 Shapes based on different polygon tracking algorithm

9.4 Conclusions

The algorithm can automatically trace the organ through the complete volume of cross sections. False contours that do not correspond to the spline shape and position can be rejected automatically from the system and can be replaced with linear interpolated contours considering as key contours those already found by the system. The BT methods used belong to the deterministic approaches and therefore there is the tendency to produce misleading results under some circumstances. To reduce that effect, data preprocessing and the gradient volume of the original CT data can be used as input to the segmentation routine. Target volume and critical structure definition is a complex and time-consuming process in radiotherapy. The complexity varies for different anatomic sites. In plan evaluation, both the physicists and radiation oncologists interact closely to subjectively identify the plan most appropriate for the individual patient. In order to reduce the investment of time and effort by the radiation oncology staff, several image analysis tools are integrated. A function that significantly accelerates the contouring process is the linear interpolation between the original key contours. The same principle can be applied for defining structures in both planar planes, sagittal and coronal. Organs with large differences in their intensities can be segmented semiautomatically. In terms of user effort, the only action required from the user is the selection of an initial point from the algorithm on the original axial slices. The complete 3D geometry of the organ will be traced automatically. Some of the common organs with high sensitivity factor and vital importance are the lungs, the spinal cord, and the trachea [14, 15, 16, 36, 37]. In addition to these organs, the external body contour can be extracted in a similar manner. The contours that are generated semiautomatically can be manipulated and modified in the same manner as those defined manually.

References

1. Adams, R., & Bischof, L. (1994). Seeded region growing. *IEEE Transaction on Pattern Analysis and Machine Learning, 16*(6), 641–647.
2. Amenta, N., Bern, M., & Kamvysselis, M. (1998). A new Voronoi-based surface reconstruction algorithm. *Annual Conference on Computer Graphics (SIGGRAPH '98), USA,* 415–421.
3. Bajaj, C., Bernardini, F., & Xu, G. (1995). Automatic reconstruction of surfaces and scalar field from 3D scans. *Annual Conference on Computer Graphics (SIGGRAPH '95), USA,* 109–118.
4. Behr, J., Choi, S. M., Großkopf, S., Hong, H., Nam, S. A., Peng, Y., Hildebrand, A., Kim, M. H., & Sakas, G. (2000). Modeling, visualization, and interaction techniques for diagnosis and treatment planning in cardiology. *Computers & Graphics, 24,* 741–753.
5. Belshi, R., Pontvert, D., Rosenwald, J. C., & Gaboriaud, G. (1997). Automatic three dimensional expansion of structures applied to determination of the clinical target volume in conformal radiotherapy. *International Journal of Radiation Oncology, Biology, Physics, 37,* 689–696.
6. Blake, A., & Isard, M. (1998). Active contours: The application of techniques from graphics, vision, control theory and statistics to visual tracking of shapes in motion. London: Springer-Verlag.
7. Boissonat, J. (1998). Shape reconstruction from planar cross sections. *Computer Vision, Graphics And Image Processing, 44,* 1–29.
8. Cohen, E., Riesenfeld, R. F., & Elber, G. (2010). *Geometric modeling with splines: An introduction.* Natick, USA: A.K. Peters Ltd.
9. Frenzel, T., Albers, D., Hohne, K. H., & Schmidt, R. (1997). Problems in medical imaging in radiation therapy. In H. U. Lemke, et al. (Eds.), *Proceedings of CARS 1997 (Computer Assisted Radiology and Surgery)* (pp. 381–387). Amsterdam: Excerpta Medica ICS 1134, Elsevier.
10. Grosskopf, S., Park, S. Y., & Kim, M. H. (1998). Segmentation of ultrasonic images by application of active contour models. *Proceedings of CARS 1998 (Computer Assisted Radiology and Surgery), Tokyo, Japan,* 871–877.
11. Grosskopf, S., Encarnação, J. L. (Referent), & Sakas, G. (Referent). (2002). Realitätsnahe Modellierung und Visualisierung dynamischer medizinischer Bilddaten mittels aktiver Konturen, aktiver Regionen und deformierbarer Modelle. Technischen Universität, Darmstadt.
12. Hoppe, H., DeRose, T., Duchamp, T., McDonald, J., & Stuetzle, W. (1992). Surface reconstruction from unorganised points. *Annual Conference on Computer Graphics (SIGGRAPH '92), USA,* 71–78.
13. Karangelis, G. (2004). *3D simulation of external beam radiotherapy.* Ph.D. thesis, University of Darmstadt, Darmstadt, Germany.
14. Karangelis, G., & Zimeras, S. (2002a). An accurate 3D segmentation method of the spinal canal applied on CT images. In *BVM 2002, Confedrence Proceedings* (pp. 366–369). Berlin, Germany: Springer-Verlag.
15. Karangelis, G., & Zimeras, S. (2002b). 3D segmentation method of the spinal cord applied on CT data. *Computer Graphics Topics, 14,* 28–29.
16. Karangelis, G., Zimeras, S., Firle, E., Wang, M., & Sakas, G. (2001). Volume definition tools for medical image applications. 4th MICCAI International Conference. In M-A. Viergever, T. Dohi, & M. Vannier (Eds.), *Lecture Notes in Computer Sciences* (Vol. 2208, pp. 1295–1297). Utrecht, Netherlands: Springer-Verlag.

17. Kass, M., Witkin, A., & Terzopoulos, D. (1987). Snakes: Active contour models. *IEEE First International Conference on Computer Vision,* 259–268.
18. Kessler, M. K., & McShan, D. L. (1994). An application for design and simulation of conformal radiation therapy. In R. A. Robb (Ed.), *Visualization in Biomedical Computing, Proceedings SPIE 2359* (pp. 474–483), Rochester, MN.
19. Kuszyk, B. S., Ney, D. R., & Fishman, E. K. (1995). The current state of the art in three dimensional oncologic imaging: An overview. *International Journal of Radiation Oncology, Biology, Physics, 33,* 1029–1039.
20. Laurent, P. J., Le Mehaute, A., & Schumaker, L. L. (1994). *Curves and surfaces in geometric design.* Natick, USA: A.K. Peters Ltd.
21. Levoy, M., et al. (1988). Display of surface from volume data. *IEEE CGA, 8.*
22. Mclnerney, T., & Terzopoulos, D. (1996). Deformable models in medical image analysis: A survey. *Medical Image Analysis, 1,* 91–108.
23. Meyers, D., Skinner, S., & Sloan, K. (1992). Surfaces from contours. *ACM Transactions on Graphics, 11,* 228–258.
24. Osorio, E. M. V. (2003). *Surface reconstruction.* Diploma thesis, Universidad EAFIT Escuela de Ingenieria departamento de Informatica y Sistemas Medellin.
25. Payne, B. A., & Toga, A. W. (1994). Surface reconstruction by multiaxial triangulation. *IEEE Computer Graphics and Applications, 14,* 28–35.
26. Pekar, V., McNutt, T. R., & Kaus, M. R. (2004). Automated model-based organ delineation for radiotherapy planning in prostatic region. *International Journal of Radiation Oncology, Biology, Physics, 60,* 973–980.
27. Raut, S., Raghuvanshi, M., Dharaskar, R., & Raut, A. (2009). Image segmentation—A state-of-art survey for prediction. *Proceedings of International Conference on Advanced Computer Control 2009* (pp. 420–424, ISBN 978-1-4244-3330-8), Singapore, Jan 2009, IEEE Computer Society.
28. Sakas, G. (1993). Interactive volume rendering of large fields. *The Visual Computer, 9,* 425–438.
29. Sakas, G., Karangelis, G., & Pommert, A. (2001). Advanced applications of volume visualization methods in medicine. In S. Stergiopoulos (Ed.), *Advanced signal processing handbook: Theory and implementation for radar, sonar, and medical imaging real-time systems.* CRC Press (ISBN 0-8493-3691-0, 2001).
30. Schmidt, R., Schiemann, T., Sclegel, W., Hohne, K. H., & Hubener, K. H. (1994). Consideration of time dose patterns in 3D treatment planning. An approach towards 4D treatment planning. *Strahlentherapie und Onkologie, 170*(5), 292–302.
31. Terzopoulos, D., & Fleischer, K. (1988). Deformable models. *The Visual Computer, 4,* 306–331.
32. Terzopoulos, D., Platt, J., Barr, A., & Fleicher, K. (1987). Elastically deformable models. *Computer Graphics, 21*(4), 205–214.
33. Turk, G., & O'Brien, J. F. (1999). Shape transformation using variational implicit functions. *Annual Conference on Computer Graphics (SIGGRAPH '99), USA.*
34. Weinstein, D. (2000). Scanline surfacing: Building separating surfaces from planar contours. *Proceeding of the 11th IEEE Visualization 2000 Conference, USA.*
35. Wells, D. M., & Niederer, J. (1998). A medical expert system approach using artificial neural networks for standardized treatment planning. *International Journal of Radiation Oncology, Biology, Physics, 41,* 173–182.
36. Zimeras, S., & Karangelis, G. (2001). Semi-automatic segmentation techniques for CT medical data. *3rd caesarism Computer Aider Medicine, Bonn, Germany.*
37. Zimeras, S., Karangelis, G., & Firle, E. (2002). *Object segmentation and shape reconstruction using computer-assisted segmentation tools* (p. 96). San Diego, USA: SPIE Medical Imaging.

Index

N. Sklavos et al. (eds.), *System-Level Design Methodologies for Telecommunication*,
DOI 10.1007/978-3-319-00663-5, © Springer International Publishing Switzerland 2014

MIX
Papier aus verantwortungsvollen Quellen
Paper from responsible sources
FSC® C105338

If you have any concerns about our products,
you can contact us on
ProductSafety@springernature.com

In case Publisher is established outside the EU,
the EU authorized representative is:
Springer Nature Customer Service Center GmbH
Europaplatz 3, 69115 Heidelberg, Germany

Printed by Libri Plureos GmbH
in Hamburg, Germany